Evolution and Man's Progress

Evolution
and Man's Progress

Edited by
HUDSON HOAGLAND *and* RALPH W. BURHOE

1962

Columbia University Press, *New York and London*

This group of essays originated in the
American Academy of Arts and Sciences
and was first published in the summer 1961 issue
of its journal *Dædalus*.

Copyright © 1961 by the American Academy of Arts and Sciences
Copyright © 1962 by Columbia University Press
Library of Congress catalog card number: 62-8741

Manufactured in the United States of America

Contents

1   HUDSON HOAGLAND AND RALPH W. BURHOE
    Introduction

6   JAMES F. CROW
    Mechanisms and Trends in Human Evolution

22   HERMANN J. MULLER
    Should We Weaken or Strengthen our Genetic Heritage?

41   *Comments on Genetic Evolution*

67   JULIAN H. STEWARD AND DEMITRI B. SHIMKIN
    Some Mechanisms of Sociocultural Evolution

88   WALTER A. ROSENBLITH
    On Some Social Consequences of Scientific
    and Technological Change

104   *Comments on Cultural Evolution*

124   B. F. SKINNER
    The Design of Cultures

137   HENRY A. MURRAY
    Unprecedented Evolutions

160   *Cultural Evolution as Viewed by Psychologists*

177   Notes on the Authors

179   Glossary

181   Conferences on Evolutionary Theory and Human Progress

HUDSON HOAGLAND AND RALPH W. BURHOE

Introduction

BECAUSE OF THEIR SPECIAL KNOWLEDGE, scientists and scholars may be in a position to see some of the future consequences and costs of current social practices before they become evident to decision makers, either in large enterprises like governments or small enterprises like families. Few have either the time or the talent to become informed about long-range implications arising from advances in science and technology. And yet in the twentieth century, as compared with previous centuries, the impact of science and technology upon our ways of living and our destiny has become paramount. The mushrooming clouds of new notions and new patterns of behavior are altering the nature and circumstances of human life more within a few years than they were altered over centuries in the past. There is no guarantee that these new circumstances are going to be stable or viable; instead, there is a probability that a civilization, whose decision makers operate in ignorance of where this onrushing current of events is taking us, may unwittingly be headed for disaster.

Of course, the future consequences of present actions and circumstances can hardly be predicted in detail by anyone. But perhaps those who are scouting the frontiers of new knowledge can envisage potential opportunities and disasters ahead of their contemporaries. In our time it has been the physicists familiar with the characteristics of nuclear weapons and the biologists aware of effects of radiation who have taken the initiative in warning the public of the dangers of nuclear testing and nuclear war. It has been the social scientists concerned with demographic problems and the biologists and other scientists familiar with the limitations on food and other necessities who have posed the problem of the population explosion.

This volume is a modest exploration of some other longer- and shorter-range problems on which some light is shed by evolutionary theory as it applies to man. This theory is found in two semi-independent branches: (a) the biological sciences, which are relevant particularly for understanding what produces the organic base of human society, and (b) the behavioral and social sciences, which are relevant particularly for understanding the various behavioral patterns insofar as they are established or propagated by social interaction. We sketch briefly the latest theories of human evolution, both in its biological or genetic and in its cultural or behavioral forms, and raise some of the practical and ethical problems with which we are faced.

A significant backdrop for this contemporary "Quo vadis?" of mankind is provided by Michael Lerner:

> It has been stressed that man is a unique animal for a great many reasons. Thus man is dominant over all other forms; he is not involved in interspecific competition, except in special cases of parasitism of himself and some of his sources of food; he has perfected methods of what may be called the Lamarckian transmission of information, that is, cultural information, methods which are much more efficient in many ways than Mendelian or biological transmission; he has a variety of nonbiological homeostatic mechanisms for preservation of steady biological and social states at his disposal; he has of course, the potentiality, if not necessarily currently the ability, of controlling the biological makeup of his near and remote descendants; he has a great variety of other properties. Above all, he is the only beast to hold conferences on his future. And, of course, the fact to remember is that the assessors of conditions under which these unique powers can be put to use and the implementers of these powers are members of his own species, and therefore subject to the same influences, prejudices, and lapses of rationality as the individuals whom he wants to manipulate. In short, the observers and the manipulators are part of the observed and manipulated. And this is the particular reason why I think we must approach our task with considerable circumspection.

The first area sketched concerns the question of what is happening and what might happen to the human genotype, that basic biological heritage which is today recognized as the foundation of human nature, the recipe for organic life, that is transmitted from generation

Introduction

to generation in the code book only recently opened up by the geneticists and biochemists. Just as the theory in physics which related mass to energy in nuclear particles allowed us for the first time to say something about the possibilities of nuclear energy, so the contemporary theories of genetics permit us for the first time to say something about the trends and potentialities of the evolution of the human gene pool.

Relevant aspects of the genetic theory are presented in the paper by James F. Crow. The essay by H. J. Muller focuses the light of these theories upon some of our current social practices and he thereby sees some dire consequences. Then he suggests some utopian possibilities if we mend our ways.

The discussants of these two papers, including some geneticists who are as well informed as the writers, point to certain doubts about the adequacy of the theory, but its major features seem to stand. They also raise doubts concerning the practical and ethical problems of following Muller's suggested program of progress. The reader can judge for himself whether our present social practices, including medical and economic care engendered by a well-meaning humanitarianism, may, because of our ignorance of their genetic consequences, be threatening the health and happiness of our descendants; whether we can do something to remedy the situation; and whether we might even produce a better breed. The importance of these matters for the future of man requires that we begin to acquaint ourselves not only with the genetic but also with the social, ethical, and all other relevant information in order to make sound judgments.

The second area sketched in this book concerns what is happening in that other major stream of human heritage which the anthropologists call "culture," the source of the language we speak and the literature we read, the technology and gadgets we employ, from agriculture and airplanes to zeros, the social structures in which we live (the family, religion, the school, economy, and government), in short, all the sources of our nature which are systematically accumulated and transmitted to us through social rather than biological channels. This kind of heritage is not appreciably shared by any other animals of this earth and, as Lerner implies, it gives man a unique capacity to adapt and develop, making him lord of all other life on earth. The transmission of the cultural heritage seems to be essentially different from and independent of the mechanisms transmitting the genotype, and the evolution of the type or pattern of culture is many times more rapid than that of the genotype.

Julian H. Steward and Demitri B. Shimkin in their paper present a comprehensive summary of what cultural anthropologists understand about the evolution of culture. A second paper on the cultural scene by Walter A. Rosenblith presents a biophysicist's view of the implications of the growth of science and the increasing coupling of science and technology in the patterns of our social structure and operations, a process that foreshadows a future more awe-inspiring than the coupling of the human mind to automatic machinery to perform man's labor—namely, the symbiosis and extension of the human brain with computer mechanisms to help man think.

As Steward and Shimkin imply, it seems that our as yet very rudimentary analyses of the tremendous complexities of culture and its changes have not and perhaps cannot give us a very secure base for seeing the future potentialities of man, and we found few discussants from among the social scientists who were willing to project a program for cultural improvement or Utopia as did Muller for biological improvement. The need felt at the conference out of which these papers arose was for greatly intensified studies of these intricate problems. The present views of the nature of cultural change, with the added vision of the vast potential changes offered by science and technology, cause us to ponder man's future in new dimensions.

The next two papers are by psychologists, who sketch some of the problems of the relation of the individual to society, problems of ordering or controlling individuals to behave in such a way that we can have a viable or workable society, and who touch on problems of human values such as freedom and love. They also illuminate further the dynamics of cultural evolution.

B. F. Skinner's paper suggests that patterns of behavior are established by social or natural reinforcements of behavior trials in a way analogous to the establishment of the genotype by natural selection. He interweaves our ways of talking about purpose and values so tightly into the fabric of scientific chains of cause and effect and so vividly does he present the powers coming out of recent behavioral science to control the individual that he arouses grave fears for

Introduction

human spontaneity and freedom in the minds of some of the discussants.

The final paper by H. A. Murray reminds us of some of the most pressing practical problems that threaten the future of man. He suggests a program, derived from his insights as a psychologist and analyst of culture, which he thinks could contribute to our well being, with specific suggestions as to how to deal with the dilemma of being either overwhelmed by nuclear war or of submitting to a distasteful, alien way of life.

The six main papers were developed for a series of three conferences on the topic "Evolutionary Theory and Human Progress," held under the auspices of the American Academy of Arts and Sciences with the support of a grant by the Carnegie Corporation of New York. Limitations of space prevent our including more than a small part of the valuable discussion among the authors and the six dozen participants invited. We are deeply indebted to the able scientists and scholars present at these conferences, for whose comments there is no room, for the invisible contribution their presence and discourse has made to the quality of what is here presented. And we are further indebted to those whose insight and wisdom steered the development of the conferences. The names of all those more immediately concerned with the conferences will be found on page 181.

The views expressed by both writers and participants are, of course, their own. No attempt has been made to arrive at a stated consensus. Since each of the conferences had its own attendance, with only a partial overlap, a general consensus of the group would not have been possible, even had it been intended.

JAMES F. CROW

Mechanisms and Trends in Human Evolution

THE MOST IMPORTANT TREND today in the mechanisms of human evolution is the radically changing pattern of birth and death rates. Until recently the death rate was high, particularly for children, while those who lived to maturity showed a high reproductive rate. Now, in most technologically advanced countries, there is a very low death rate and a voluntarily low birth rate. The most important consequence is the effect on world population. The birth rate has fallen rapidly, but in most areas not so rapidly as the death rate, so that population grows at a faster rate than ever. If we can avoid a thermonuclear war, then surely the most pressing problem for men in the next century will be one of sheer numbers.

The principal concern of this paper, however, is not the absolute numbers of the population, but the trends in the kinds of persons who compose it. The guiding force in biological evolution is selection. Selection implies that some types of individuals leave more descendants than others do, whether because of their greater ability to survive or their greater fertility. Mutation provides the variant forms that are the raw materials of evolution. The gene-shuffling process we call Mendelian inheritance is a mechanism for arranging these variants in endless combinations. The structure of the population may permit some isolation of subgroups with possible evolutionary consequences; but the general direction of the evolution of the population is determined mainly by selection. Those genes that on the average enhance the fitness of their carriers will increase in numbers in later generations. In this connection, "fitness" is used in a strictly Darwinian sense as a measure of actual survival and of reproduction, and not (as J. B. S. Haldane has cautioned) fitness for "football, industry, music, self-government, or any other activity."[1]

Differential Survival and Reproduction

First of all, we must realize that selection depends on differential mortality and fertility. No genetic change will ensue unless individuals having different genetic makeups contribute in different proportions to future generations. At the crudest level, we can measure the extent to which different individuals of one generation are differentially represented by children in the next generation, irrespective of the causes of the differences. As a measure of genetic selection, this is clearly unsatisfactory, for many such differences are purely accidents of time and space; automobile wrecks and thermonuclear bombs kill tall and short people, indiscriminately. To some extent, differential survival and reproduction depend on an individual's characteristics, such as his health; but health is partly determined by early environment.

The human birth rate is an example of the effect of early selection. A newborn child of intermediate weight has a greater chance of surviving for the next few months than a larger or smaller child has. To this extent there is selection for birth weight; but birth weight is certainly not determined by the gene makeup alone: the intrauterine environmental factors are also of great importance. Even if traits were entirely determined by heredity, this would not necessarily mean that the population is readily changed by selection. To continue with the same example of weight at birth: a gene that causes a slight increase in weight will be disadvantageous in an infant that is otherwise too large, but advantageous in a small child. A size-determining gene, therefore, is subject to contradictory pressures in selection, and its frequency in the general population accordingly does not change much, even though the population is subject to intense selection.

The conclusion is that selection alone is not always capable of effecting a genetic change in a population. Some traits are largely unresponsive to selection; if we select a group of seven-footers as parents, we will not necessarily have a population of giants in the next generation. Animal and plant breeders use a statistical concept called "heritability" to describe the effects of selection on a population. A trait that has a high heritability is readily changed by selection; a trait that remains only slightly changed, or not changed at all, has a low heritability.[2] There may be several reasons for a low heritability: (1) the trait may be determined environmentally, rather than genetically; (2) on the other hand, it may be largely or even completely determined by the genes, and it will still have a low heritability if the gene frequencies are in some sort of selectively balanced state, or if there are complex interdependencies between different genes or between genes and environment.

In discussing the trends in human evolution, let us consider the effect of existing patterns of births and deaths on the genetic makeup of the population, and what the effect of various possible eugenic proposals might be. The answers to such questions are unknown, except in broad outline. As a preliminary question, let us return to the amount of potential selection inherent in the pattern of births and deaths, whether or not it is genetically effective, so as to enable us to set some limits on the amount of selection that could occur.

Has Human Selection Nearly Stopped?

In 1850 about one-fourth of all the children born in the United States died before the age of five years. The 1959 death rates are such that 63 years would have passed before this proportion of the population would have died.[3] There has been a similar though less spectacular fall in the birth rates in the last century. Does this mean that natural selection has been virtually eliminated as an evolutionary agency in the human species?

If every individual had the same number of descendants, whatever that number, there would be no selection. As emphasized earlier, selection depends on the differential contribution of different individuals—that is, on *variability* in fitness. The appropriate measure of this variability is what I shall call the "Index of Opportunity for Selection." This index shows the rate at which the genetic composition of the population would change if all the survival and reproductive differences were directly reflected in changes in gene frequencies. For the statistically minded, this index is the square of the coefficient of variability in the number of progeny per parent.[4]

Mechanisms and Trends

It is revealing to divide the index into two components, one related to deaths, the other to births. The index of opportunity for selection by death has greatly decreased in the last century, as would be expected from the falling death rate. If the mortality between birth and the average age of reproduction is 50 percent, then the index is 1.00; if the death rate is 10 percent, it is 0.11. These two death rates correspond roughly to those of the late nineteenth century and the recent past, so that in two generations the index of selection by death has dropped to about one-tenth of its former value.

Clearly, the low death rate has reduced the opportunity for selection by death. Whether the actual genetieal selection has been as drastically reduced is another question. In earlier times, most deaths resulted from infectious diseases, and it is likely that not many deaths were appreciably influenced by the individual's genotype. Today, a much larger fraction of all pre-adult deaths occur from recognizably genetic causes. It is probably true, therefore, that the amount of genetically effective selection by death has not been as drastically reduced as the drop in the index would suggest. It is nevertheless true that the amount of opportunity for changes in genes because of differential mortality is greatly reduced.

During this same period, on the other hand, the index of selection because of differential fertility has actually risen in many technologically advanced countries. The data in Table 1 show the values for several selected populations. The pattern is consistent, striking, and perhaps surprising. With the fall in the birth rate, the index becomes greater rather than less. The highly fertile populations are uniformly fertile, whereas the changing pattern of birth rates has somehow produced more variability in the number of children. It may well be that very recent changes in the United States may have reversed this change somewhat, as contraceptive practices have spread more uniformly through the population; but the general trend is very clear. Just as the decrease in the index of selection by death does not mean a corresponding decrease in genetically selective deaths, an increase in the index of selection by birth does not imply a necessary increase in genetic change. Many factors that determine the size of families in a population in which this is largely a voluntary decision are not genetically determined.

The statement made in the above paragraph that recent trends in the United States are the reverse of the long-time trend can be documented. The reversal is more striking than I had expected. Norman Ryder has kindly called my attention to some newer data (reported by him in *Demographic and Economic Change in Developed Countries*, Princeton University Press, 1960, page 125). In the cohort of women reaching age 45-49 in 1960 the mean has risen to 2.6 and the index has dropped to .68; in the cohort one decade later, although

TABLE 1

Index of Opportunity for Selection because of Differences in Fertility in Various Populations *

Population	Mean number of children	Index
Rural Quebec	9.9	.20
Hutterite	9.0	.17
Gold Coast, 1945	6.5	.23
New South Wales, 1898-1902	6.2	.42
United States, born 1839	5.5	.23
United States, born 1866	3.0	.64
United States, aged 45-49 in 1910	3.9	.78
United States, aged 45-49 in 1950	2.3	1.14
Ramah Navaho Indians	2.1	1.57

*The figures are based on the total numbers of children born to women who had survived to the end of the reproductive period. For the sources of these data, see J. F. Crow, "Selection," in *Symposium on Methodology in Human Genetics* (in press).

the families are not all completed, the mean is over 3 and the index is less than .5. So it appears that recent trends toward uniformity of family size along with a slight increase in the average have reduced the index to the level of much earlier times.

The important conclusion is this: the amount of differential reproduction that now exists, a differential of which we are hardly conscious and which we do not ordinarily regard as a social burden, would, if applied to a highly heritable trait, be such as to make possible a rapid change in that trait. The loss of opportunity for

Mechanisms and Trends

selection because of a lower death rate is approximately compensated for by a greater opportunity, one that is inherent in the pattern of birth rates.

Prenatal Mortality

Although most prenatal mortality is very early and is therefore not a large cause of human misery, it is probably a major factor in selection for some human genes. In man, the total embryonic death rate is unknown, since there is no accurate measure of the number of conceptions; but in cattle, sheep, mice, swine, and rabbits, it averages some 30 or 40 percent. It now appears likely that two factors, not previously thought important, are substantial contributors to this death rate.

One factor is chromosome abnormalities. The disease of mongolism has been found to depend on the presence in triplicate (rather than duplicate) of one of the small chromosomes, so that there are 47 chromosomes instead of the regular 46. This irregularity occurs with a frequency of about one in 700 births. The corresponding deficiency type with only one chromosome is probably more common, if we can judge from data on Drosophila, and it probably leads to embryonic death. Aberrations because of similar errors on the part of three other chromosomes have recently been discovered. The remaining chromosomes presumably make mistakes, too, and probably most of these result in embryonic death, for the cases so far discovered, despite their being attributable to some of the smallest chromosomes, are grossly abnormal children. If we consider the large number of chromosomes and the various kinds of errors and rearrangements that are possible, it is likely that these chromosomal abnormalities are a substantial cause of embryonic mortality.

The other newly realized factor in embryonic death is maternal-foetal incompatibility. Apparently, an embryo of blood group A or B with a group O mother has a risk of some 10 percent of dying as an embryo, presumably from antigen-antibody reactions between embryo and mother. On the other hand, in combinations in which no such incompatibilities exist, the group O embryos, and presumably the other homozygous types, are found in deficient numbers at birth and therefore must have died as embryos. Altogether, it has been estimated that about 6 percent of all fertilized eggs die before birth owing to the effects of this single gene locus. If this analysis is correct, the ABO blood-group genes are the largest known single cause of genetic mortality.

If the other blood group factors (MNS, Rh, Kell, Lewis, and several others) have effects of comparable magnitude, then these few genes must account for a substantial part, perhaps a majority, of all embryonic deaths. Nothing is known about most of the other genetic polymorphisms, such as different types of serum proteins. Perhaps they too are maintained by differential embryonic death.

We should not conclude that all embryonic mortality is owing to blood-group genes and to chromosome aberrations. In cattle, swine, and guinea pigs, there is a definite rise in embryonic mortality when the parents are consanguineous. Such an association of embryonic death with inbreeding shows that at least some such deaths must be attributable to recessive genes, and presumably the human organism is the same in this respect. Finally, it is clear that not all embryonic death is because of the embryo's genotype. There is a marked increase in embryonic death in swine and guinea pigs when the *mother* is inbred. Thus, embryonic survival depends to a large extent on the mother's health and vigor, which in turn depends on her genes and her environment. I know of no information that would indicate any striking change in the pattern of embryonic deaths in recent years. This would seem to be the one part of the picture of human natural selection that has not changed radically.

Postnatal Death

The spectacular drop in postnatal death rates has been in the main because of a greater control over infectious diseases. One after another of the epidemic diseases have been eliminated as major causes of death, either by prevention or cures. Haldane has argued that for the past several thousand years the major selective factor in human survival has been resistance to disease. As man changed from a nomadic hunting life to one of more densely populated agricultural and manufacturing communities, the danger of death from wild animals became less, while the danger from communicable disease must have increased.

Mechanisms and Trends

Today these diseases have virtually disappeared from much of the world. Moreover, this change, when measured against the long years of man's history, has been virtually instantaneous. This means that whatever genetic mechanisms of resistance to disease were developed during our thousands of generations of contact with the diseases, these mechanisms are still with us. Resistance to disease in part depends on general health and vigor, and to this extent the genes selected for disease resistance in the past are still valuable. But Haldane has suggested that resistance to infectious diseases in many cases involves highly specific mechanisms that are of no use in other contexts and may in fact be harmful.

An example is the sickle-cell anemia gene (common in Africa), because in some manner it confers a measure of resistance to one type of malaria. Despite the fact that a double dose of the gene causes a severe anemia that would nearly always be fatal in a rigorous environment, the gene is retained because in a single dose it leads to increased malaria resistance. Today many populations, well on the way to freedom from malaria, thanks to modern sanitation and insecticides, still carry a gene which still exacts its price in anemia deaths with no longer any compensating benefit from malaria resistance. There is the possibility that several genes whose present function is obscure are relics of former disease resistance mechanisms in which they were some way involved, or were selected for other less obvious reasons that are no longer relevant.

The Effect of a Changed Environment

In the history of most evolving species a change in the environment is almost always bad. The reason is that, through natural selection in the past, the organism has acquired a set of genes that are well adapted to this particular environment. A change in the environment is almost certain to necessitate some gene replacements. Thus, much of evolution is spent keeping up with what, from the standpoint of any one organism, is a steadily deteriorating environment, the most serious and rapidly changing aspect of which is the improvement of competing species through their own evolution. Man, however, is unique in that most changes in the environment are of his own doing. Therefore, the environment instead of deteriorating is getting better, at least for most genotypes. Consider, for example, the precipitous drop in childhood death rate, with only a slight compensatory rise in adult death from such diseases as cancer.

I suspect that in the majority of cases a changed environment ameliorates rather than obliterates the harmful effect of a mutant gene. A gene that is harmful in one environment is usually so in others as well, though not necessarily to the same extent. For example, the science of blood transfusion has decreased the severity of hemophilia but has by no means rendered it innocuous. Insulin has greatly helped to control diabetes, but the insulin-treated diabetic would be still better off if he had not had the disease at all.

If the effect of an environmental improvement is to lessen rather than remove completely the harm caused by a gene, the effect is one of postponement rather than prevention of the harmful effect. Consider a lethal mutant gene that causes the death of its bearer. Suppose that a new drug reduces the probability of death to 10 percent. The gene will now persist in the population for an average of ten generations before it is eliminated from the population. By a system of mutation cost-accounting, in which ten persons exposed to a one-in-ten risk of death are equal to one person with a certainty of death, the drug has not helped except by postponement.

As human beings, we are primarily interested, not in the effect of a gene on fitness *per se*, but in its associated effects on health, happiness, intelligence, and other aspects of human well-being. From the standpoint of long-range human welfare, the most beneficial kind of environmental advance is one that reduces the amount of suffering and unhappiness caused by a mutant gene by a greater degree than the increase in fitness. On the other hand, an environmental change that, for example, increases the fertility of persons with a severe or painful disease without a corresponding decrease in the amount of suffering caused by the disease will in the long run cause an increase in human misery.

Another consideration that must inevitably enter into any such discussion as this is the cost of the environmental improvement. For it must be remembered that any such improvement must be a permanent one, continued from this time on. No one need regret the inability of the human body to manufacture some of our needed vitamins; they are cheap and easily available in the food supply. Nor

Mechanisms and Trends

is a genetically conditioned susceptibility to smallpox any great problem. There is no reason to think that this disease will not remain under effective and relatively cheap and simple environmental control.

On the other hand, a person who has a genetic defect that is corrected only by expensive surgery or by repeated blood transfusions throughout a lifetime might well consider having a smaller number of children than he otherwise would, so that this cost, measured either in economic hardship or human suffering and frustration, might be correspondingly less to the next generation. Furthermore, a genetic defect may be costly to others than the affected person— his family, his community, society as a whole—if, for example, he requires institutional care.

Mutation

So much has been written recently on this subject that any interested person can find a number of articles that are authoritative or easy to read, and in some cases both. I shall be content with only a few remarks. Many abnormal genes owe their frequency in the population to the "pressure" of recurrent mutation. In each generation new mutants arise, to be eliminated later, often very inefficiently, by natural selection. Eventually, these processes will come to balance, though in man it is likely that the conditions determining the balance are changing so fast that the population never has time to reach an equilibrium.

It is clear that under this circumstance the harmful effect of mutation is proportional to the number of new mutants that arise. In fact, if the metric chosen is simply Darwinian fitness, the harm done to the population is exactly equal to the total mutation rate, as was first shown by Haldane and then independently by H. J. Muller.[5] A major problem for future research is how the effect can be measured in more tangible terms of human happiness or satisfaction. If the harmful genes are maintained by a balance between opposing selective forces, the effect of an increase in the mutation rate is no longer easily measured, even in units of Darwinian fitness; but the direction of the effect is the same, though less. Despite disagreement in details among geneticists, the conclusion that an increased mutation rate is harmful is universally accepted.

I would maintain that the ideal mutation rate for the human population, now and in the near future, is zero. It might be argued that mutation is needed for the maintenance of genetic variability for future evolution, and ultimately this is true; but for the time being and a very long time in the future, the reservoir of existing genes, which are to some extent pretested by the very fact of their continued existence, can supply any variability that man needs for his evolution. If mutation were to stop entirely, we should probably not know it for thousands of years, except by a reduced frequency of such diseases as hemophilia and muscular dystrophy. Meanwhile, it is clear that the present rate is too high for our immediate welfare, and anything that can be done to decrease it, or at least keep it from increasing, is to the good.

This brings up the question of the radiation hazard. There is reason to believe that the present levels of radiation from natural sources, medical and dental uses, and from the military and industrial uses of nuclear energy are probably responsible for only a minority of human mutations—though we must agree that the evidence is largely circumstantial and from experimental animals. Certainly, the amounts of radiation should be reduced whenever possible. But I should like to urge that more attention be given to possible chemical mutagens. In our complex chemical society, it is quite possible that some widely used compounds are highly mutagenic. And of course, should a way be found to lower the "spontaneous" rate, so much the better.

In summary, there is enough genetic variability for man's evolution in the foreseeable future without the introduction of new mutations. Surely, a combination of genes and environment that admits of a range of physique from pigmies to giants, of mental abilities from an idiot to Shakespeare, Newton, and Mozart, and of ethical standards from Himmler to Schweitzer provides as wide a range of variability as we could need—although one recalls Haldane's remark that, if we want a race of angels, we would have to obtain new mutations, both for the wings and for the moral excellence.

I should point out that the existing range of variability in a population does not limit the future possibilities that might arise by a recombination and selection of existing genes. If, for example, Mozart's constitution had contained the maximum concentration of genes favoring musical genius of anyone yet known, this does not by any means imply that his was the best genotype that could have been constructed from the pool of genes available in a large population.

Population Structure

Another trend in recent human evolution has been the increasing amalgamation of previously isolated sub-populations. The increased interdependence of various geographical regions and increased ease of travel are rapidly destroying the earlier pattern of a world divided into relatively small isolates, each undergoing a separate biological and cultural evolution. In a small population, all the people are likely to be somewhat related. Therefore, this trend toward isolate-breaking has an immediately beneficial biological effect through a reduction in the amount of inbreeding. To this extent the benefits of hybrid vigor are added to whatever effects accrue from cultural amalgamation.

Whether the outbreeding effect is substantial or small is difficult to determine. Any attempt to study this is hampered by the confounding effects of environmental changes accompanying the presumed genetic effect. This is especially true for interracial marriages, for the hybrid population is likely to find itself in a social environment different from that of either parent group.

Sewall Wright has argued that for maximum evolutionary opportunity a subdivided population is best. When isolates disappear, some opportunity to test different gene combinations and (probably more important) to develop unique cultures is lost. Yet the steady breaking-down of isolates is going on, and the trend is likely to continue. I would suggest that some of the immediate genetic effects (for example, less inbreeding) are beneficial, and that the genetic variability will be conserved about as well either way. To me, the social consequences of the disappearance of some cultures is more serious. I believe that in the long run the best solution is not to oppose cultural and genetic fusions, but to work for a society which tolerates the greatest amount of individuality, whether this is genetic or acquired.

JAMES F. CROW

Birth Selection instead of Death Selection

As mentioned earlier, the deleterious effects of many (I think most) mutant genes are diluted and postponed by an improved environment, rather than obliterated. Despite the fact that many of the deaths in the past were accidental or caused by diseases that are not now relevant, some of them were genetically selective in eliminating mutants that otherwise would now be causing harm. To the extent that these deaths no longer occur or are reduced in number, the genetic makeup of the population is deteriorating. We may not be conscious of this, for the environment is improving too rapidly, but the damage is still there to be reckoned with by any society that is conscious of its genetic future. The question is: can we use birth selection in an effective and socially acceptable way to compensate for decreased death selection?

To suggest to a person whose hereditary disease has been cured or repaired that he have no children at all may be (except for severe and highly heritable conditions) an undue imposition. But to reduce the number by half perhaps would make little difference to his happiness yet effect a substantial genetic change, if the practice were widespread. The reduction of future human misery by the detection of normal persons who carry hidden harmful genes is effective for some diseases and can become more so as knowledge increases. To prevent the reproduction of all persons with genetic feeble-mindedness (phenylketonuria) would change the incidence of disease less than 0.5 percent in the next generation. But if each heterozygous carrier had only half as many children as he would otherwise have, or if sperm from another person were used when the husband is a carrier, this would reduce the abnormal-gene frequency by 50 percent, and the disease incidence for the next generation would be greatly reduced.

There is a limit to this possibility, though. Probably we each carry some half-dozen detectably harmful genes. Eventually, a system would be required by which those persons who knew they had an exceptionally large number of disease-producing genes would voluntarily reduce the number of children they produced.

It is generally agreed that such diseases as muscular dystrophy, hemophilia, and phenylketonuria should be controlled. Most people would agree that parents who are likely to have children with such traits should be informed. Hereditary counseling is surely in order, and is already widely accepted. I hope that society will revise its notions about therapeutic abortion and artificial insemination, wherever genetic diseases are involved.

On the other hand, society is not agreed, nor is it likely to agree in the near future, about what if anything is to be done about such quantitative, usually multigenic traits as intelligence. A few words are in order about the extent to which such changes could be made, if there were a conscious selection of some sort. The history of animal breeding has shown that substantial changes can be made by selection for quantitative traits when these have not been strongly selected in the past. The ability of dog breeders to produce bizarre forms is apparent to everyone. It is also striking that the milk production of an average cow now is far better than the best some years ago. On the other hand, selection for increased viability and fertility in livestock has been considerably less successful.

This probably means that it would be easier, by selection, to change the intellectual or other aptitudes of the population than to change the incidence of disabling diseases or sterility. This is not to say that there has not been some selection for intelligence in the past, but it has surely been much less intense than that for fertility, for example.

Since society owes so much to a small minority of intellectual leaders, a change in the proportion of gifted children would probably confer a much larger benefit on society than would a corresponding increase in the population average. These potential leaders would probably produce enough change in cultural and other environmental influences to be worth considerably more than the contribution of their genotypes to the genetic average. It has frequently been suggested that when artificial insemination is used, because of sterility or genetic disease in the husband, the donors might be selected from men of outstanding intellectual or artistic achievement. Were this widely practiced, I believe that the occasionally highly gifted children, though probably a small proportion of all the children produced, might still be a most important addition to society.

It is important to realize that even a very intense selection may make only very slow changes in the gene makeup of the population, since many of the traits of greatest importance have a low heritability. Similarly, any dysgenic effects of an accumulation of mutations because of medical advances or an excessive reproductivity of the genetically less well endowed, also act very slowly. Therefore, while we need not rush into hasty or ill-considered solutions, it is time to start discussing the problem. We must remember that natural selection has been cruel, blundering, inefficient, and lacking in foresight. It has no criterion of excellence except the capacity to leave descendants. It is indifferent as to whether living is a rich and beautiful experience or one of total misery. Post reproductive ages are of no consequence, except in so far as older parents and grandparents aid in the survival of the young.

Selection under individual human control, on the other hand, could be greatly different. Its means are birth selection, not death selection. It can make use of scientific knowledge such as biochemical tests of carriers of harmful genes, or genetic knowledge of relatives. It can have foresight. It can have criteria of health, intelligence, or happiness—not just survival and fertility. And it can make use of the various scientific and technological advances (sperm storage, egg transplant, and other more distant prospects) as they are discovered.

We must remember also that the decision is not as to whether man should influence his own evoution. He is already doing this by his revolutionary changes in the environment, by medical advances, by the invention of contraceptives. The issue is not *whether* he is influencing his evolution, but in what direction.

REFERENCES

1 J. B. S. Haldane, "Parental and Fraternal Correlations in Fitness," *Annals of Eugenics*, 1949, *14*: 288.
2 For discussions of this concept, see D. S. Falconer, *Introduction to Quantitative Genetics* (London: Oliver and Boyd, 1960); and I. M. Lerner, *The Genetic Basis of Selection* (New York: Wiley, 1958).
3 *Statistical Bulletin* of the Metropolitan Life Insurance Company, March 1960.
4 This was originally called the "Index of Total Selection," but I gladly accept G. G. Simpson's suggestion that it be called the "Index of Opportunity for Selection." The Index is defined as the variance of the number of offspring

per individual divided by the square of the mean number. The children are counted at the same age as the parents, and those who die before the reproductive age are counted as leaving zero descendants. For example, if half the persons have 3 children and half have 1, the mean number is 2, the variance is 1, and the index is ¼. If half have 6 and half have 2, the mean is 4, the variance is 4, and the index is the same, ¼, as it should be, for these two populations have the same evolutionary consequences except for total numbers. For a derivation, justification, and discussion of this index see J. F. Crow, "Some Possibilities for Measuring Selection Intensities in Man," *Human Biology*, 1958, *30*: 1-13.

5 H. J. Muller, "Our Load of Mutations," *American Journal of Human Genetics*, 1950, *2*: 111.

HERMANN J. MULLER

Should We Weaken or Strengthen our Genetic Heritage?

Selection in Man, Past and Present

Before the development of culture, biological evolution had already laid the genetic basis in our ancestors for a number of more or less distinctive faculties of body and mind. Among these were erect posture with its associated greater adroitness of the hands and greater facility in using them, a comparatively high level of curiosity and of general intelligence, a social, cooperative disposition, and, more peculiarly, symbolizing propensities and urges to vocalize diversely, to imitate, and to communicate. The use of these faculties resulted in the gradual accumulation of primitive culture.

Reciprocally, culture opened up modes of life that gave increasing opportunity for the effective exercise of these very faculties. In so doing, culture afforded an increasing relative advantage in the struggle for existence to those individuals, families, and small groups of families in whom the genetic bases of these faculties were better developed. Thus, in submen and primitive men, over the course of hundreds of thousands of years, there must have been a positive feedback by which cultural evolution aided biological evolution and *vice versa*.

This article is an abridged version of "The Guidance of Human Evolution," in *Perspectives in Biology and Medicine*, 1959, 3: 1-43, and in Sol Tax and Charles Callender (editors), *Evolution of Man: Mind, Culture and Society*, Chicago, University of Chicago Press, 1960; it is reprinted here with the permission of the editors and publishers. The writer's discussion of the same topic appears in Sol Tax and Charles Callender (editors), *Issues in Evolution: The University of Chicago Centennial Discussions (Evolution after Darwin*, vol. 3), Chicago, University of Chicago Press, 1960.

However, as culture advanced to the stage of larger and fewer groups, the intergroup natural selection necessarily became increasingly inefficient, and as the social relations within these groups resulted in a more effective extension of aid to individuals and families in need of it, the intragroup natural selection also slackened off. Today it is evident that these two processes are rapidly approaching their limit: that of a worldwide, *de facto* socialized community in which everyone is helped to live according to his need and to reproduce according to his greed, or lack of foresight, skill, or scruple. Thus we cannot extrapolate from the past to the future and say that our cultural advance will continue to result in biological betterment. On the contrary, a negative feedback from culture to genetics has now set in. In other words, the saving of lives for reproduction by ever more efficient medical and other technical and sociological aids tends to result in an increasing accumulation of detrimental mutations that occur at random. These must adversely affect health, intellect, powers of appreciation and expression, and even the genetic bases of our cooperative disposition itself. At the same time, the disappearance of the semi-isolating barriers between small groups removes the main basis (often pointed out by Sewall Wright) for evolutionary experiments that may result in break-throughs.

In fact, it seems not unlikely that in respect to the human faculties of the highest group importance—such as the neuronal equipment conducive to integrated understanding, foresight, scrupulousness, humility, regard for others, and self-sacrifice—modern cultural conditions may actually lead to a lower rate of reproduction on the part of their possessors than the rate of those with the opposite attributes. Is it not too often true today, when birth control is available, that those persons are likely to have the largest retinue of children, whether legitimate or otherwise, who are most lacking in perspective, or are dominated by superstitious taboos, or are unduly egotistical, or heedless of others' needs, or shiftless, or bungling in techniques? These considerations raise the possibility that a much faster acting and more serious cause of genetic deterioration than the accumulation of detrimental mutations accurring in the wake of relaxed selection is an actual reversal of selection in regard to those psychological traits that are of the highest social importance. Objective data are badly needed on this question.

The relaxation of selection that has been mentioned, as well as its postulated reversal, has been made possible chiefly by the fact that under modern culture the damage done by social as well as other defects of individuals is increasingly borne by society as a whole. This in itself is a great step forward. However, its potential consequences for the biological evolution of man have not been considered seriously enough, even though Darwin himself called attention to them.

On the other hand, it is indisputable that, as man's control over matter advances, more and more of his bodily structure and functioning can be amended to advantage, and even replaced by artificial means. But a given endowment is better lost than retained only if all the following three conditions hold. First, the artificial substitute should be more effective and dependable than the natural endowment. Second, the net burden to the community involved in the maintenance and operation of the substitute should, for a given return, be less for the artificial substitute than for the natural endowment. Third, the maintenance of the natural endowment as a supplement to the man-made contrivance should be more trouble than it is worth before its lapse can be considered justifiable.

Today, of course, no attempt is made to assess these balances when procedures are instituted which, in helping individuals, may contribute to the relaxation of selection in given directions. The real issue here is not whether society should in this way help the individuals themselves to live better, but whether the acts of society should be so ordered as actually to facilitate the perpetuation of defective genetic equipment into later generations. For, by so doing, we give with one hand while taking away with the other.

Future Consequences of Our Genetic "Laissez Faire"

Let us try to make a rough quantitative appraisal of the consequences of the relaxation of selection alone. On the average, the counterpressure of typically acting selection, consisting of the elimination of individuals with excess detrimental genes, almost exactly equals the pressure of mutation in producing these genes. There is evidence from more than one direction that in man, at least one person in five, or twenty percent, carries a detrimental gene which arose in the immediately preceding generation. There are several lines of evidence for this. One is based on the frequency of mutations found in such different organisms as mice and Drosophila. Another is derived from the frequencies with which, on inbreeding, human beings arise that are defective enough to die before maturity. In each of these cases certain assumptions that seem reasonable have to be made. Thus, in the first case, one must assume a similar number of genes, and in the second case, a similar amount of dominance for genes in the different species. Despite these assumptions, the answers come out remarkably close to one another—that is, something like twenty percent, regardless of which method is used. One in five carries one additional defective gene besides perhaps scores of defective ones derived from the previous generation. Now, this same proportion of individuals—one in five—is usually prevented by genetic effects from surviving to maturity or (if surviving) from reproducing. This dying out keeps the balance, so that the total load of mutations does not go on increasing after this rate of the elimination of mutant genes has been obtained.

This equilibrium holds only when a population is living under conditions that have long prevailed. Modern techniques are so efficacious that, used to the full, they might today (as judged by recent statistics on deaths) be able to save until the reproductive age some nine-tenths of the otherwise genetically doomed twenty percent. Let us assume this to be the case, and that those saved for reproduction usually take about as much advantage of the opportunity for it as the rest do. In the next generation there would then be eighteen percent who carried along those defects that would have failed to be transmitted in the primitive or equilibrium population, plus another twenty percent (slightly overlapping the eighteen percent) who had the most recently arisen defects.

At this rate, if the effectiveness of the techniques does not diminish substantially as their job grows, there would be, after somewhat less than eight generations, or 240 years, an accumulation of about 100 "genetic deaths" (scattered, however, over many future generations) per 100 persons then living, in addition to the regular "load of mutations" that any population would ordinarily carry. It can be estimated (on the supposition that human mutation rates are like those in mice) that this amount of increase in the load is about the same as would be brought about by an acute exposure of all the parents of one generation to 200r of gamma radiation (a situation similar to that at Hiroshima) or by a chronic, low-dose-rate exposure of each of the eight generations to 100r (as a result, for instance, of industrial or medical usages).

This result sounds worse than it is because most of the mutant genes would cause only a slight amount of damage in their usual heterozygous condition (i.e., when received from only one parent). The regular load of perhaps some scores of significantly detrimental genes per individual—nearly all in heterozygous condition, but each a potential cause of some far future genetic death—gives rise to a total risk of genetic extinction, as measured under primitive conditions, that is something like (but probably exceeding) twenty percent for any *given* individual. It is evident that this addition of just one more detrimental gene to his load would usually increase the individual's own risk of extinction by only some tenths of one percent. The process of genetic decline would therefore be exceedingly slow, thanks to the innumerable factors of safety that are built into our systems. The decline, in fact, would for a very long time consist mainly in a reduction in these factors of safety. For our organism has all sorts of ways of arriving at a result and of bolstering the set of balanced processes which some people call the healthy state.

Let us next suppose that this sparing of genetic deaths by the aid of technology is to continue indefinitely at the assumed rate, a rate at which a genetic defect, on the average, subjects a person to only a tenth as much risk as it would if he were living under primitive conditions. Eventually, after some tens of thousands of years, a new equilibrium would be reached at which the load of mutations would be about ten times as large as at present. At that time, as many extinctions as mutations would again be occurring. If we are to keep to our chosen figure for a mutation rate, there would be one extinction for every five individuals, or twenty percent. The frequency of genetic deaths would accordingly have returned to the level it had in primitive times—far above that now prevailing, in spite of all technological efforts.

At the same time, the average individual of that period, carrying ten times today's genetic load, would, if tested under primitive conditions, be found to be no longer subject to a risk of extinction of only twenty percent, but to one of 200 percent. This means that he would usually carry twice as much defect as would suffice to eliminate him. Man would therefore have become entirely dependent on the techniques of his higher civilization. Yet, even with these techniques, he would be subject to as high an incidence of genetic misfortunes as had afflicted him in primitive times—that is, his weaknesses would have caught up with him. Along with this, he would be devoting a far greater proportion of his time and efforts than now to supporting the social burden of the whole community in caring for these weaknesses in the population at large.

Although the further improvement of techniques would result in a long-continued pushing back of the time of attainment of equilibrium, there would be in consequence an even greater increase in the frequency of detrimental genes. Suppose, for example, that a saving was finally achieved of all except one in a hundred of the genetic lives which under primitive conditions would have been lost. This situation would lead toward an equilibrium which, though attained only after some hundreds of thousands of years, resulted in a one hundred-fold increase in the frequency of detrimental mutant genes, as compared with their present frequency.

At that time, and for a very long time before it, each individual would be endowed by nature with a unique assortment of many hundreds of cryptic as well as conspicuous inherent defects, most of them individually rare. He would therefore constitute a pitiful combination of special cases. He would have to be given a superlatively well-chosen combination of treatments, training, and artificial substitutes, merely to survive. The task of ministering to infirmities would consume all the energy that society could muster, leaving no surplus for general cultural purposes. Yet, even at that stage, *most* of man's genes would still be fairly normal, so that new mutations, always arising at the rate of some twenty percent or more in every generation, would continue to plague him still further.

Long before such an "advanced" stage of the genetic *cul-de-sac* were reached, however, this medical utopia would probably have been subjected to such great strains as to have thrown men back toward more primitive ways of life. Many would then have found themselves incapable of such ways. To be sure, the difficulty then might in one sense be said to be "self-rectifying," as one of our military manuals, commenting on the mutations induced by radiation, has maintained. Yet so late and so forced a "rectification" would be likely to cause the loss of much that man had previously gained, and might even bring about his complete collapse.

The Application of Ethics to Parenthood

Luckily, however, our present culture has brought us knowledge of this situation, and of evolution as a whole. Much as in the case of the dangerous techniques placed in our hands by physics, chemistry, the knowledge of how to deplete our resources, and the use of mass media for thought control, we see that there can be no effective renunciation of our powers, if we would retain the benefits of civilization. Instead, we must meet all these difficulties by mustering greater foresight and introducing more social motivation in the use of our knowledge and skills and in the further extension of our understanding and self-control.

In the realm of genetics, this means that, in the light of an understanding of how biological evolution works, men actuated by a more far-reaching sense of responsibility to their successors will come to extend their social awareness and their motivation not only to their contemporaries but also to the next and succeeding generations. In the course of learning (as they *must*) to control their numbers, it is to be expected that they will increasingly come to recognize that the chief objective in bringing children into the world is not the glorification of the parents or ancestors by the mere act of having children, but the well-being of the children themselves and, through them, of subsequent generations in general. Becoming better educated in genetics, they will become aware that in some measure their gifts as well as their failings and difficulties—physical, intellectual, and temperamental—have genetic bases. Moreover, social approval or disapproval will then be accorded them, according to whether or not they take these matters into account in their decisions regarding reproduction.

A new kind of pride in reproduction will arise when they realize that, if they are burdened with more than the average share of defects, their most valuable services will lie in the exercise of a greater than average degree of control over the multiplication of their own genes and in contributing to the community in other than directly genetic ways. Conversely, those persons more fortunately endowed will feel it their obligation to reproduce to more than the average extent.

In these matters, it can become an accepted and valued practice to seek advice (though not dictation), even as is done in matters of individual health. Such advice is especially likely to be widely sought because everyone has many minor, and some more serious, inherent imperfections, and it is the total balance of these and of his gifts that should count. For this reason, moreover, the answer need seldom be an all-or-none one. Thus, someone might be advised that it would be better if he had only three-fourths as many children as the average person.

Now, the same kind of change in attitude toward reproduction as is needed to insure the preservation of our genetic heritage provides in addition the necessary basis for its improvement. If once it is accepted that the function of reproduction is to produce children who are as happy, healthy, and capable as possible, then it will be only natural for people to wish each new generation to represent a genetic advance, if possible, over the preceding one, rather than just a holding of the line. And they will become impatient at confining themselves to old-fashioned methods, if more promising ones for attaining this end are available. As the individualistic outlook regarding procreation fades, more efficacious means of working toward this goal will recommend themselves. In time, children with genetic difficulties may even come to be resentful toward parents who had not used measures calculated to give them a better heritage. Influenced in advance by this anticipation and also by the desire for community approval in general, even the less idealistic of the parental generation will tend increasingly to follow the genetic practices most likely to result in highly endowed children.

Artificial Insemination as an Instrument in Ethical Parenthood

The most effective method of positive genetic selection at present feasible in man is, of course, artificial insemination. Many thousands of people have already been begotten in the United States by this procedure in cases of male infertility—a considerable portion of them by the use of sperm from donors other than the husband (while in the rest the sperm from the husband himself was introduced, often after it had been concentrated by centrifugation). Under such circumstances, the couples concerned would nearly always be receptive to the suggestion that they turn their exigency to their credit by having as well endowed children as possible.

It is highly important that the genetic paternities of the children thus produced be properly recorded. Unfortunately, however, most physicians who today conduct inseminations with the sperm of donors do so furtively, almost as though everyone concerned were guilty. Moreover, they fail to seek the best genetic material and fail to provide genetic records. On the contrary, it should be recognized that the couple concerned, as well as the physician, have performed a service to mankind that merits not disgrace but honor. With such an outlook, even before it is *generally* held, both physician and couple would be armed with better incentives to take genetic considerations into account. They would be encouraged to make the best possible use of such a chance to engender the most precious thing we know: a worthy human being.

Fortunately, a very high intensity of selection is made possible by this method, because of the large excess of spermatozoa that each individual usually gives rise to. Recently, the further technique of freezing spermatozoa to very low temperatures has been introduced into practice and, thanks to the accumulation of sperm thus made possible from any given individual, the frequency of successful conceptions has been raised considerably above that following artificial insemination of the more usual type, or even natural insemination. These deep-frozen spermatozoa can be stored virtually indefinitely without deterioration. Thus, a considerable supply can in time be gathered from a chosen donor and preserved for any desired length of time. For purposes of selection, there would be enormous advantages in such cases in postponing the use of most of this supply until, say, twenty years after the donor's decease. In retrospect, and after the fading away of much of the emotionally based pressure and prejudice, both pro and con, that had arisen out of the donor's personal relations, much less biased judgements could be made concerning his actual merits as well as his shortcomings. During the "probationary period," moreover, a limited but significant amount of data regarding his genetic potentialities might be forthcoming, from a consideration of his descendants and other relatives who had arisen in the meantime.

Such a procedure would also be of considerable psychological advantage, for it would eliminate that bane of present-day gynecologists who practice artificial insemination—the fear of detrimental personal relations arising between the woman concerned and the donor, if either learns the other's identity. The source of a motivation for possible jealous reactions on the part of the husband would at the same time be greatly diminished. Moreover, both members of the couple, as well as all other persons, would be much more likely to recognize the donor's exceptional worth, a worth that would usually put him out of a class with possible living competitors. Finally, the chief present objections to having the identity of the donor known would be removed. Instead, all interested persons, once the practice had become an accepted one, would wish the relations involved to be entirely above-board.

It remains true that some long-entrenched attitudes, especially the feelings of proprietary rights and prerogatives about one's own germinal material, supported by misplaced egotism, will have to yield to some extent. This feeling does not represent a natural instinct, however, since there are primitive peoples yet alive who do not even have the concept of biological fatherhood, and other peoples, such as the Hawaiians, who, although having it, readily and without their parental relations being affected, adopt, confer, or exchange infants. I have known white people in Hawaii who had become Hawaiianized to this extent, and who introduced me to their Chinese child, or their child of native Hawaiian ancestry, along with their white children, in quite a natural and matter-of-course way.

To more than balance the necessary weakening of this time-worn vanity in regard to one's stirps, other feelings will tend to develop that are of equal or greater potency. Among them will be a justifiable pride in an accomplishment of a far more exacting and laudable kind than that of procreation: that of having made children of especially high endowment possible through the use of one's rational powers and social awareness, and of having brought these children up as one's own. Correspondingly, deep attachments to these children will develop, and a justifiable sense of identification.

These new reproductive mores will of course come into being only very gradually in society as a whole. Just as our economic and political system is inevitably, although too slowly, being modified to fit our present technological capabilities of large-scale automatic production, despite the fervor with which men try to cling to their ancient preconceptions of how business and government should operate, so too on the biological side of human affairs the time-honored notions of how reproduction should be managed will gradually give way before the technological progress that is opening up new and ever more promising possibilities.

Practices that today are confined to couples afflicted with sterility will be increasingly taken up by people desiring at one and the same time to bestow on themselves children with a maximal chance of being highly endowed, and also to make an exemplary contribution to humanity. Making some sacrifice in the matter of traditional feelings of vainglory about the idiosyncrasies of their personal stirps and braving the censure of the old-timers, they will form a growing vanguard that will increasingly feel more than repaid by the day-by-day manifestations of their solid achievements, as well as by the profound realization of the value of the service they are rendering. Thus there will be no sudden break with tradition, and in the same family children of chance and children of choice will grow up side by side in mutually helpful association. "Nothing succeeds like success," and the successes in these instances will often be outstanding.

Technical Possibilities in the Offing

We are surely just around the corner from other advances in artificial techniques concerned with reproduction that might extend the possibilities of positive selection much further. For example, there have as yet been only a few abortive attempts to cultivate either male or female germ cells outside the body. An energetic program of research on the subject might succeed within a few years in enabling spermatogonia, at least, to be multiplied indefinitely *in vitro* and to be induced, when desired, to undergo the processes of maturation into spermatozoa. If this could be done, then spermatogonia instead of spermatozoa might more practicably be preserved in the deep-frozen state, a technique that has already proved successful with some types of somatic cells. For later, at any chosen dates, the spermatogonia could be multiplied to any desired extent and caused to mature, so as to furnish an unlimited supply of mature spermatozoa from an originally small amount of material derived from any given donor. Only our present superstitious attitudes prevent such research from being actively pursued today.

As for the female germ cells, means are already known by which the multiple release of mature eggs can be readily effected within the female with the aid of a pituitary hormone. Comparatively little research would be required to develop suitable methods of flushing out these eggs from the female reproductive tract, to be fertilized *in vitro* with chosen sperm, and then implanted in selected female hosts at the appropriate stage of their reproductive cycle. This procedure would be parallel to artificial insemination. It would permit the multiple distribution of eggs of a highly selected female into diverse recipient females, yet when so desired it would enable the child to be derived on its paternal side from the recipient's husband. Possibly, too, techniques involving mature eggs could be combined with deep freezing to allow indefinitely prolonged storage.

We cannot go so far afield here as to discuss the diverse possibilities of still further technical achievements in reproduction. One such achievement would be a parthenogenesis of a type not involving meiosis. By its means, mankind might be enabled to reap the benefit of that alternation of asexual reproduction (for reliably multiplying types of tested worth) with sexual reproduction (for trying ever new combinations) which has been so advantageous in some other groups of organisms. Later generations will look with amazement at the shamefully small amount of research now being carried on to open up such possibilities, even though for decades specialists have realized that efforts in these directions are likely to prove unusually fruitful.

Genetic Considerations

It should be admitted freely that no one, except to a pitifully limited degree, can predict the results of any matings in such heterogeneous material as an existing human population. With rare exceptions, we certainly cannot proceed in humans according to the principles governing *simple* Mendelian differences. For the multifactorial (sometimes called "polygenic") basis of most phenotypic differences in human and other natural, cross-breeding populations is undisputed —especially when they concern traits of importance, and have been therefore long subject to natural selection—even though some gene differences do, of course, produce major effects, along with many more that produce minor ones. We can seldom hope in practice to know just what genes are concerned in a given case. Neither could we say just what the effects of given genes in untried combinations would be. Nor can we control or predict what combination a given zygote will have. In addition, it is often impossible to discriminate between genetically and environmentally based effects.

Even though all this is true, nevertheless, in the over-all picture, selection carried out on a purely empirical basis does work effectively. Even primitive man was able to do much better than nature usually did, in effecting changes in his domestic races in the directions he sought at a pace much faster than that at which species ordinarily change when under natural conditions of breeding. This was so because his manner of selecting was less haphazard and more single-tracked than nature's. The selection of domestic forms practiced by man in the present century has been much more rapid and effective still, as for instance in the milk yield of cattle, for it has been greatly helped by knowledge of the genetic principles at work even though gene identification is seldom feasible. This knowledge has shown, among other things, the importance of better assessing an individual's genetic potentialities by the study of his progeny or other relatives.

It should also be recognized that in all selectional work in which the potentialities of the individual genes in further combinations are not known, some chosen for multiplication will later turn out (when in homozygous condition or in given groupings) to have undesirable or even lethal effects. This happens also in natural matings. To be sure, the careful statistical analysis of progeny could reduce such events to an even lower than natural frequency. But in any case, a general progress will take place in the direction of selection anyway, and it can, especially in the earlier stages, be very rapid. Then, in later stages, the delicate task of screening for the most dependable and precisely suited genes can be better carried out, after the general level has been raised and a plentiful supply of promising genetic material has become available. In the meantime, the favorable phenotypic effects of the early rise in the level can be enormous.

Toward What Should We Aim?

Thus the important thing, especially in the earlier stages of any course of selection in man, is the *kind of trend* that is instituted. As to this, the madness of the racists has taught the world by terrible object lessons the dangers of egotism, ethnocentrism, and particularism in general. One of the main antidotes is a better, more vivid teaching of evolution, with its emphasis on the fundamental unity of man and the overriding importance of the species as a whole, and with its underscoring of the paramount values that men the world over have already come to cherish deeply.

Among the qualities of man most generally valued are a genuine warmth of fellow feeling and a cooperative disposition, a depth and breadth of intellectual capacity, moral courage and integrity, an appreciation of nature and of art, and an aptness of expression and of communication. It has been through the exercise of these faculties that man has raised himself to his present estate. Yet most people, if they are honest, will grant that these qualities have never been in oversupply, and that, as our culture advances, we can make increasingly good use of a higher level of them. This higher level should be striven for by both cultural and genetic means.

It is evident that the distribution of the relative strengths of different drives that was most appropriate to the success of people when they were divided into many small, nearly autonomous groups is far from that most suitable for men organized into a vast society engaged in scientific and technological advances, mechanized production, transportation and communication, predominantly common interests, and, at least in intention, rational, democratically guided decisions. Surely, in this society, most of us could do better if by nature as well as by training we had lesser tendencies to quick anger, blinding fear, strong jealousy, and self-deceiving egotism. On the other hand, we need a strengthening and extension of the tendencies toward kindliness, affection, and fellow feeling in general, especially toward those personally far removed from us. As regards other affective traits, there is much room for broadening and deepening our capacity to appreciate both natural and man-made constructions, to interpret with fuller empathy the expressions of others, to create ever richer combinations of our own impressions, and to communicate them more adequately to others.

Another direction in which advances are needed is in those traits of character that lead to independence of judgment and its necessary complement, intellectual honesty. We need to strengthen the drive to see things through to as near the bottom as possible, and also the drive to coordinate the elements rationally. Just as important are the will and ability to take fair criticism with good grace and, further, to search and criticize ourselves until we recognize and discard, if need be publicly, judgments based on wishful or faulty thinking or on defective data. A great deal of this may be taught, but there also seem to be great inborn differences in the facility and degree with which such emotion-fraught mental operations are learned and in the strength of feeling behind them.

To turn now to more purely intellectual matters, it is obvious that tomorrow's world makes desirable not only a lively curiosity but also a much greater capacity for analysis, for quantitative procedures, for integrative operations, and for imaginative creation. With more and more of the daily grind taken over by automation, the human being will be increasingly freed for higher mental jobs. Yet most of our population today would be by nature ill-adapted for such activities, even if they had the desire to pursue them. They also miss the thrilling awareness of the vastly enlarged possibilities that a more rational use of these and other powers could open up to everyone. If, in this scientifically and technologically based mass society, men are not to be mere cogs in their work and pawns in their play, they must have deeper and broader vision, as well as a more virile, broadly based comradeliness. Then their machines and their science can give them an increasing freedom for further achievement and a further savoring of the bounties of our expanding universe, instead of a deeper enslavement in routine.

Education everywhere is increasingly striving for all these objectives, and the public at large is giving ever greater recognition to their importance in bringing up children. Thus, in the adoption of genetic methods also, it will be natural to direct their use toward

Our Genetic Heritage

these same goals. In other words, unless men sank into the hands of mad or ignorant dictators, there would be no danger that in the long run they would fail to recognize and seek their fundamental values.

At the same time, especially interested groups will surely endeavor to promote the diverse abilities and proclivities of specific types as well. These will include not only intensified aptitudes of a more purely intellectual nature but also those making possible a more far-reaching and poignant appreciation of the varied kinds of experiences that life may offer. Moreover, as all these resources of mankind grow richer in their genetic basis, they will increasingly be combined to give more of the population many of their benefits at once. For observation shows that these faculties are not antagonistic but, rather, mutually enhancing. No one need fear, however, that there will be a likelihood of men's really salutary diversities becoming wiped out, so long as selection is conducted in a democratic way. And later, in a wiser, kindlier age than ours, men may more safely and calmly consider how this spice of life, variety, can be turned to even better account.

Finally, too, it will be feasible to strive for real progress in what is called the physical side: to better the genetic foundations of health, vigor, and longevity; to reduce the need for sleep; to bring the induction of sedation and stimulation under more effective voluntary control; and to develop increasing physical tolerances and aptitudes in general.

As to the question of how far we should go in all these directions, it is to be observed that in many physical respects there are optimal degrees of development for given types of organisms, beyond which other functions tend to be too much interfered with. So probably the ape would have said, "You can't, you *shouldn't* increase the size of our cranium more than 10 percent." Yet these optima are seldom absolute, for there are often ways of breaking through the seeming limits by means of novel developments that open up new methods for solving the old problems. These lines of inquiry are so advanced, however, that we may leave them to the more competent minds of the future to tackle. Our horizon of the future is today very near to us and very limited. Meanwhile, we at least have the assurance that there must be enormous stretches beyond that near horizon. For, in the opposite direction, that of the past, we have, even within the lifetime of some persons still living, at last glimpsed the grand panorama of the three to five billion years of evolution already elapsed. We have come to know what seeming miracles the plasticity of protoplasm (or rather, of nucleic acid) is ultimately capable of performing.

Cultural and Biological Evolution Must Pull Together

It is easy to sit back smugly in the conceit that we have now reached almost the acme of biological evolution and that, except perhaps for eventually bestowing on everyone the genetic advantages already enjoyed by the most favored, we can hereafter confine our advances to cultural evolution, including the manipulation of things outside our own genetic constitutions. It is true that cultural evolution, in this broad sense, is far more diversified, rapid, and explosive, both figuratively and literally, than biological evolution can be. Moreover, along with the increasing understanding of and mastery over physicochemical forces that our expansion in space and our advances in control over physical energy imply, there will surely be corresponding advances in the physiological, neuropsychological and social realms. Yet all this does not mean that genetic advances beyond the stage represented by the happiest possible combinations of the best endowed of present-day humanity would be either supererogatory, unimportant, or relatively limited.

If the comparatively small genetic difference between apes and men is really as important as it seems to be in determining men's amenability to profit by culture and to contribute to it, why would not beings as far beyond present-day men, genetically, as we are beyond the apes be inordinately better suited still for exploiting the benefits of culture? Do we hastily made-over apes really believe that, having attained our present makeshift form, further steps of this kind are to be despised? Our imaginations are woefully limited if we cannot see that, genetically as well as culturally, we have by our recent turning of an evolutionary corner set our feet on a road that stretches far out before us into the hazy distance.

It is true that, by utilizing our present genetic basis, culture alone has carried us very far, that it can carry us very much farther still and, wisely developed, can give every man a fitting place under the sun. It is also true that, even with human aid, biological progress must be far slower than that of culture. But the total advance is not the sum of the two. It is more like the product or even the exponential. And even as our own culture could not mean very much to the most superior ape, so the culture of a mere million years from now will be so rich and advanced in its potentialities of experience and accomplishment that in it we with our genetic constitution of today would be like imbeciles in a palace.

We should recognize that the genetic changes desirable for us in the future would not in the main be replacements for cultural devices but the very opposite. They would be a means of better gearing together the biological and the cultural, of making still more out of our culturally enhanced propensities, and of more effectively advancing our culture. As we have seen, this evolutionary trend actually began with the advent of man (or even before it), although he was unaware of it as a long-term phenomenon and therefore allowed the further development of his culture to interfere with it. Now that we understand this matter better, we may again carry this trend forward, this time consciously and with ever longer foresight.

Only after we humans have advanced considerably toward the higher level to which the rough-and-ready empirical methods now available can raise us, will we be in a position to make firmer, more definite plans, envisaging longer-range possibilities. Only then, when we have developed superior intelligence, more comprehensive knowledge, and greater cooperativeness, as a result of a cultural combined with a genetic advance, can we expect to reach a workable degree of agreement on these plans. Only then can we begin to use more exact methods and to coordinate them better. It is too early now for blueprints.

If we are to preserve that self-determination which is an essential feature of human intelligence, success, and happiness, our individual actions in the realm of genetics must be steps based on our own personal judgments and inclinations. They should be as voluntary as our other major decisions of life that also concern society, and like those other decisions they will take into account the best knowledge and advice available. Although these decisions are all conditioned by the mores about us, these mores can to some extent be specifically shaped and channeled by our own distinctive personalities. The immediate job, then, is to make a start at getting this genetic "Operation Bootstrap" incorporated into our mores, by precept and, where feasible, by example. But we must remember that the highest values to be sought in it are in essence those so long proclaimed but seldom actualized: wisdom and brotherhood, that is, the pursuit of "the true and the good." When it is understood that the genetic method offers simply an additional but indispensable approach toward this ancient ideal, then our voluntary genetic efforts, scattered and disjointed though they now must be, will tend in a common direction.

There are sure to be powerful attempts to pull in diverse directions, in genetic as in other matters, but we need not be afraid of this. The diversities will tend to enrich the genetic background and increase the resources available for recombination. These partial attempts can then be judged by their fruits, and these fruits, where sound, will be added to our bounty.

It seems highly unlikely that in a world-wide society at an advanced level of culture and technology, founded on the recognition of universal brotherhood, such diversities would proceed so far and for so long as again to split humanity on this shrunken planet into semi-isolated groups and that these groups would thenceforth undergo an increasing divergence from one another. It is because man is potentially master of all trades that he has succeeded. And if his culture is to continue to evolve indefinitely, he must retain this esssential plasticity and versatility, and with it the feeling that all men are at bottom of his own kind.

And so, I believe, not only our cultural but also our biological evolution will go on and on, to new, undreamed of heights, each of these two means reenforcing the other, and again with a positive feedback, but with an enormously more effective one than hitherto.

Comments on Genetic Evolution

THE COMMENTATORS on genetic evolution are: Robert S. Morison, Director of Medical and Natural Sciences, the Rockefeller Foundation; Garrett Hardin, Professor of Biology, the University of California (Santa Barbara); J. Paul Scott, Chairman, the Division of Behavior Studies, the Roscoe B. Jackson Memorial Laboratory, Bar Harbor, Maine; Lawrence K. Frank, Professor of Sociology, Brandeis University; Ernst Mayr, Professor of Zoology, Harvard University; Theodosius Dobzhansky, Professor of Zoology, Columbia University; James F. Crow (see below, p. 177); Hermann J. Muller (see below, page 177); I. Michael Lerner, Professor of Genetics, the University of California (Berkeley); Ralph W. Gerard, Professor and Director of Laboratories, Mental Health Research Institute, the University of Michigan; and George G. Simpson, Professor of Paleontology, Museum of Comparative Zoology, Harvard University; Donald H. Fleming, Professor of History, Harvard University.

R. S. Morison: I must confess to some vague, ill-formed and largely instinctive misgivings about the idea of directed evolution. They have to do with the problem of saving the future from the constraints put on it by the limited vision of those who would plan it. There have always been those who feel that they have a clear idea of what man and his society ought to be like and they have devised means for molding them closer to the idealized images. Characteristically, and doubtless necessarily, these images have always been more limited than the actual results later brought about by undirected cultural evolution. The thing that has saved man from his limited visions in the past has been the difficulty of devising suitable means for reaching them.

COMMENTS

At its deepest level, the problem is not one of differentiating the bad from the good visions, although at any given point in time it is up to all of us to try. The real point is to recognize that all such visions are in the long run inadequate. To cite a biological analogy, all known visions of man and his future suffer from the evolutionary defect of overspecialization. From the biological standpoint, man, as Huxley delights in pointing out, is one of the least specialized of all organisms. This is one of the reasons he has survived for so long. But the cultures man has made for himself have all ultimately failed because of their overspecialization. In this sense, it is perhaps fortunate that, in general, cultures have had a restricted geographical coverage. As our common culture gradually extends to include our entire gene pool (a desirable development in many ways), we must give increasing attention to the dangers of overspecialization, for, in this case, failure of the culture will mean failure of the entire human race.

Professor Muller's recommendations for guided evolution raise the possibility of combining cultural evolution with biological evolution, and in doing so they necessarily raise the specter of overspecialization in a particularly severe form. One's uneasiness is the more acute because he argues so eloquently and so convincingly for the probable effectiveness of the biological procedures involved. It now seems as if we stand on the threshold of doing what man has always wanted to do—make himself over closer to the heart's desire.

Certainly, the case made for guiding evolution so as to maximize intelligence, robust physical health, and brotherliness is most persuasive. So in many ways was the case made by Plato for his republic. But are the problems of values really so simple? As a matter of pure common sense, there seems no room for dissent from those who advocate some consciously guided method of reducing the number of clearly deleterious genes in the reproducing stream. Both Professor Crow and Professor Muller clearly demonstrate that we can no longer rely on natural selection to do this for us. But we may soon find that the doctrine of "clearly deleterious" is no easier to apply in practice than is Holmes's doctrine of clear and present danger in a realm of free speech.

Even at that, it is probably much easier to agree on minimizing the "bad" than on maximizing the "good." It may indeed turn out that modern biology, by giving us the concept of balanced polymorphism, has laid a sound scientific foundation for a system of morals and ethics for which the Greeks found sanction only in their instinctive liking for the golden mean and the doctrine, "Nothing too much." Western man seems to have arrived at his concept of a pluralistic society, in which no one way of life is considered absolutely better than several others, largely because he recoiled in horror at the numbers of heretics who would have to be killed if he persisted in his search for the single "best" way of reaching salvation. The concept of polymorphism gives him a much better reason for cherishing pluralism, for it makes part of the demonstrable order of nature. It would be a pity if, just at the moment when genetics has given us this highly sophisticated notion, we should turn aside to use that science to pursue a single-minded, idealized vision of what man might become. For, in the very act of defining the good, we may limit the possibilities for goodness.

Garrett Hardin: "Man is condemned at every moment to invent man," the existentialist Jean-Paul Sartre has said. In the biological context, Sartre's aphorism has been made inescapably true by the advent of "death control" and the development of the population problem. Whether man shall direct his own evolution is no longer in question. His only options are these: (1) to do so consciously or unconsciously; (2) to minimize or maximize the role of chance; (3) to control individual actions on a directive basis, or merely by statistical biasing mechanisms; and (4) to set up one system (of government and selection), or many. "Every one knows" that we are moving closer and closer to "one world." This knowledge is part of the mythology of our time. But is it true? It matters a great deal whether it is or not, because, as Sewall Wright has emphasized, evolution in a many-world complex is quite different and richer in possibilities than is evolution in a one-world system. This has been recognized by Crow, but somewhat nostalgically, as though preserving or attaining a many-world system were no longer possible.

Will the future produce one world or many? Perhaps the history of economic thought gives us a clue. We no longer argue in terms of capitalism versus socialism, but rather we think in terms of a "mixed economy," in which some economic functions are handled socialistically, others individualistically. The argument of one world versus many will probably be similarly disposed of by the development of a "mixed community." Certainly, the problem of "the atom" must be handled on a one-world basis; but that does not preclude handling other problems on a different basis. If we can seriously entertain the idea of many cultural-and-genetic worlds, the conceptual and practical freedom so gained may lead us to discover desirable possibilities scarcely conceivable in a one-world system. It should be the task of geneticists, psychiatrists, and theoretical anthropologists (if such people exist) to explore these possibilities at the conceptual level, without being hampered by sociological presuppositions created by the mythology of the moment.

J. P. Scott: The attached diagram gives a convenient summary of the subdivisions of genetics, arranged according to levels of organization, which in turn correspond to certain major fields of science. From this we can see that genetics is one of the most basic of the biological sciences and therefore occupies a key position. However, the very fact of its being basic implies that direct genetic effects are often far removed from the most important phenomena—population, social and behavioral problems of man. Behavioral genetics is barely coming of age, and social (or societal) genetics is a field which has been almost untouched except for the attempts of the eugenicists to apply some primitive ideas of selection to human beings. Enough information is now being accumulated regarding animal societies so that we can now begin to speak of this subject in a more realistic way.

THE SUBDIVISIONS OF GENETICS

Level of Organization	*Unit of Organization*	*Subdivisions*
Ecological	Population	Population genetics
Societal	Society	Social genetics
Behavioral or psychological	Organism	Behavior genetics
Physiological	Organs, tissues, etc.	Physiological genetics
Cellular	Cells, chromosomes	Mendelian genetics
Molecular	Gene	Chemical genetics

As will be noticed, this diagram omits the field of developmental genetics which touches on all levels of organization. Any individual starts at the lowest level, from which all others are developed. This re-emphasizes what should be a truism applying to any phenotype, that behavior is not inherited but developed under the influence of all sorts of environmental as well as genetic factors. In the higher animals it goes on differentiating throughout a lifetime. Because of the enormous complexity of interaction which takes place under these circumstances, John L. Fuller and William R. Thompson have emphasized the principle of noncongruence between genotype and phenotype. This means that in a large number and perhaps a majority of cases there is no one-to-one correspondence between gene and character. I need not dwell on the complications this imposes on any program for selection. I think we can all agree that in anatomical structure man is a polymorphic species, and I believe that this same idea of polymorphism can be applied to behavior as well. Data regarding the wolf (*Canis lupus*, the presumed ancestor of our most intimately related domestic animal, the dog) indicate that this species is also polymorphic. This is not true of all social species, and we can hypothesize that the condition is associated with highly social species occupying dominant ecological positions, and that polymorphism not only does no harm to the species but may be of actual advantage. In a social group of wolves, it may be an advantage to have both timid and confident animals, each of which is useful under different conditions. Be this as it may, one of the obvious genetic consequences of polymorphism is that a high proportion of individuals in a population must be heterozygous, provided there is no assortive mating.

I now turn to a genetic phenomenon first observed by R. C. Tryon in selecting rats for their capacity for maze running. He obtained a bright and a dull strain by selection, but when he crossbred them, he discovered that the range of variation of the offspring of the first generation included both parent strains, and that those of the second generation were no more variable than those of the first. At first this looked merely like an interesting genetic anomaly with a variety of possible explanations. Since then this kind of inheritance has turned up again and again in experiments on behavior genetics, and I shall present an example in the inheritance of the capacity to develop playful aggressiveness in dogs.

The explanation of this trait seems to be that offspring of a first generation are genetically located near a threshold, so that it is

COMMENTS

readily possible for these animals to develop either type of parental behavior. Since there has been a general evolutionary tendency toward the development of the capacity for variability in behavior as a superior means for adaptation, this suggests that the heterozygous condition of this trait is actually the most desirable from the point of view of adaptive capacities. It follows that, while true breeding strains can be produced by selection, these are actually inferior to a heterozygous population. If correct, this imposes still another limitation on the improvement of any population by selection.

One of our most pressing human problems is the reduction of needless destructive and aggressive behavior, so that the process of cultural change and evolution may proceed in a constructive way. As increasing numbers of animal societies are studied under natural conditions, it becomes apparent that many species (if not all) develop highly peaceable and cooperative relations within their own group. A. Murie, who studied a pack of Alaskan wolves, saw little if any fighting within the group, and much evidence of cooperative hunting and feeding behavior. Many other examples could be given, but one of the most recent and spectacular is I. Devore and S. L. Washburn's studies of baboons on the South African plains. The male baboons combine to defend the group against possible predators and form a threat which most predators prefer not to face. They have a strict order of dominance within their group but show almost no overt violence or real bodily injury. Females in estus usually consort first with the most dominant male, but pass to the company of other males without evidence of jealousy or fighting. The basic organization of the group resides in the fact that females with young infants are found in the vicinity of the biggest male, apparently because this is the safest spot.

Contrast this behavior with that of baboons in the London zoo, as reported by S. Zuckerman. In this socially disorganized group of strange adults, the males fought one another to the death, fought over the possession of females, and sometimes literally pulled one another to pieces. We may conclude that this species of social animals has evolved the capacity to develop highly cooperative and peaceful behavior within the normal social environment (which has evolved along with the behavior), but that it also has the capacity to develop extremely destructive behavior under conditions of social disorganization. The same basic capacities may result in two very different end results.

We used to think that destructive human behavior was the result of man's primitive nature. In a sense this is true, but these examples of behavior in organized and disorganized animal societies indicate that constructive and cooperative behavior are primitive in the evolutionary sense. This raises the basic question of man's genetic limitations with regard to social change. Man's evolution has reached a point at which cultural change takes place at a very rapid rate and largely independent of biological change. We have many historical indications that such changes are limited by man's basic biological nature and that certain kinds of social environments become unbearable. If we can judge from our studies of animal societies, these basic genetic limitations rest on the patterns of social behavior peculiar to a species, and we may state that a social environment cannot distort or suppress these basic patterns of behavior in any major way.

Since heredity is one of the most basic factors affecting behavior, we might naively conclude that this is a way to produce major changes and improvements in human behavior and that the obvious method to use is selection. I have already pointed out some of the defects inherent in such a program, but can illustrate this even more graphically. The dog has been subjected to all the mechanisms of genetic change, including selection, inbreeding, random drift, etc., during 10,000 years or so of domestication; yet a survey of dog breeds reveals that the general patterns of behavior in any breed are essentially those of the ancestral wolf. These may be exaggerated or partially suppressed, but nothing has been basically changed. In fact, these basic patterns are relatively similar to those found in the whole family of *Canidae*, even including foxes. What, then, is the role of the geneticist? One highly valuable function is to assess the results of social change, as Crow has done in his paper. Again, may I illustrate with a canine example. In the show breeds of dogs, the most common method of breeding is outbreeding, combined with selection for an arbitrary type which may change radically from time to time. In addition, a grand champion male may be bred to hundreds of females within a breed population. From an immediately practical standpoint, the dog breeder can most rapidly achieve results by the method described above. Genetically, the system is an almost perfect method for perpetuating and disseminating recessive genes. Our experiments on inbreeding purebred dogs indicate that these animals are a veritable storehouse of injurious recessives affecting viability and fertility.

Now, the present trend of human mating is toward outbreeding, not only in relation to families, but also in relation to communities and world populations. As with dogs, this should result in the preservation of recessive mutations. I would also like to suggest that the practice of artificial insemination from one supposedly superior male has considerable genetic dangers. If this person has a perfect set of genes, the results would be excellent, but if he has even a few injurious recessives, these could be widespread throughout the population instead of remaining in a small proportion, as they now do. One may argue that this risk would be counterbalanced by the spread of "good" genes, but I have already pointed out the lack of congruence between genes and complex behavioral characters. In simple terms, it is the combination that counts, not a particular gene.

The final function of the geneticist—and this must be in cooperation with other scientists—is to attempt to define basic human nature so that we can predict in advance the consequences of cultural and environmental changes.

L. K. Frank: Are we likely today to be misled by a Darwinian conception of human development that regards any social practice that limits biological disaster as dysgenic, and, if so, has the history of human civilization been a prolonged mistake? Surely, the first shelters, the beginning of agriculture and animal husbandry, the practice of surgery and medicine, were all interference with the elimination of all but the fittest to survive. Where and when does human intelligence and the ability to cope with and transform the environment become dysgenic?

If we are to consider the future of the human species, with its highly significant but relatively short past cultural experience, we should be aware of the danger of judging biological processes with purely contemporary social-economic, political, religious, legal, and esthetic criteria. The characteristics which early in man's history were essential to his survival, such as aggression and emotional reactions, may be recognized as of the greatest significance at that time, but today and tomorrow they may be evaluated differently. At one time war was regarded as a way of eliminating the less fit. Are efforts to prevent wars dysgenic?

May we ask, are the considered theories of geneticists today as solid and unassailable as they seem to be in these presentations? Are we extrapolating from fruit flies and mice and other infrahuman organisms to man? While the medical sciences have advanced our understanding of the human organism by the study of infrahuman organisms, our understanding of human behavior and of personality development and expression have not been similarly furthered, because man lives in a symbolic cultural world of meaning and goals.

Perhaps the most important question we must raise is, where does man with his mammalian ancestry differ from other organisms, not only in his observable functions and processes but also in his human potentialities? He has, we know, a large and more effective capacity for maintaining his internal environment dynamically stable under a variety of external and internal conditions, a prolonged infancy and adolescence which make possible more and longer learning, a deferred puberty which also keeps him juvenile and plastic, capable of a slower orientation to adult living, and he has a long period of reproductive capacity.

It took millions of years of evolutionary development from fish, through amphibians, reptiles, birds, and then the succession of mammals to develop female heat so that the female would be ready to accept and attract the male for copulation at the time, and only at the time, when she ovulates and is ready to be impregnated. This was a genetic trait that was essential to survival, we may assume. But the human female is apparently a mutation who has lost the gene for heat. Was this a genetic deterioration, or an emancipation from a strict biological control of sex that has made possible many basic human advances, especially the evolution of a human sexuality and all that accompanies it, so that human sexuality is not confined to copulation for reproduction?

The genetic source of human potentialities cannot be questioned, since that is the only origin of the living organism; but what we know about the genetic inheritance of man, as the two foregoing papers acknowledge, is scanty, and the criteria used in evaluating the variety of organism-personalities in which the genetic background is revealed do not seem either adequate or wholly credible when we view them in the light of human history and the diversity of cultures. When man emerged as *Homo sapiens*, he found a world already well populated with various organisms, each of which had already established and occupied its life zones. The competitive exclusion principle, which Garrett Hardin has recently emphasized, finds its most dramatic exemplification in man, who, unable to compete with other organisms on their terms for the possession of a life zone, developed his culture as a way of creating a human, coexistent world for human living—that is, a human life zone.

Ernst Mayr: We must try to overcome as rapidly as possible the lag of information from one group in our society to the others. For instance, it is not sufficiently stressed in the literature that there has been a considerable conceptual revolution in biology—the replacement of typological thinking by population thinking.

Now, just what do I mean by that? Population thinking means that we think in statistical terms. If, for instance, an evolutionist speaks of a tall race as compared to a short race, he simply means that the arithmetical mean of the one is greater than that of the other. He does not mean that every individual of the tall race is tall and that every individual of the short race is short. This would seem very elementary and simple. Yet in the discussions between evolutionists and geneticists, on the one hand, and nonbiologists, on the other, we often find that when the evolutionist uses a term like "superior" or "more intelligent," the nonbiologist thinks these adjectives are used in an absolute and typological sense. The evolutionist means nothing of the sort. All he means is that there is a difference in the arithmetical average, a statistical difference. Such a difference may be of vital importance in any problem of selection. Let us keep constantly in mind, therefore, that whatever value statement an evolutionist makes is made in a statistical, populational sense.

A second misconception I have found in some of the anthropological literature is the assumption that cultural transmission and cultural evolution have completely replaced biological evolution. Dr. Crow's demonstration of the many selection pressures still operating clearly shows how wrong this is. Biological and cultural evolution are superimposed on each other, they are not alternatives.

Genetic Evolution

Considering how often the word "fitness" is misunderstood among nonbiologists, I want to return to this term. The evolutionist when mentioning fitness refers to a very specific type of fitness, Darwinian fitness: this is the contribution to the gene pool of the next generation. Any gene or genotype that more than replaces itself from one generation to the next has "superior" fitness. A gene or genotype that exactly maintains itself in the gene pool has a fitness of unity. This is a strictly operational definition, and as soon as we understand this meaning of fitness, we realize that natural selection is still operating in man and is responsible for genetic changes from one generation to the next.

It is important to emphasize that there exists virtually no trait whose phenotype is not strongly influenced by both the environment and the genetic basis, and by both at the same time. This is particularly true of behavior traits. The calculations of the mathematical evolutionists have shown that even traits with only a 1 percent genetic component of heritability, if they have a selective advantage, may in the long run lead to considerable evolutionary change.

It must be emphasized that technical difficulties will prevent the genetic analysis in man of all characters except a few oligogenic traits, like the blood groups and the metabolic and some other genetic diseases. We have no adequate genetic analysis of a single polygenic trait in man, and one must assume that the characteristics that are particularly important to mankind—the components of character, intelligence, and so forth—are highly polygenic. It is important, however, that we do not derive conclusions from the study of oligogenic traits and apply them uncritically to highly polygenic traits.

Th. Dobzhansky: I should like to emphasize a point Professor Crow mentioned only briefly. He quite correctly states, "A changed direction of selection is almost certain to have left us with a heritage of a number of now harmful genes." However, usefulness and harmfulness are not the intrinsic properties of a variant gene; genes are useful, neutral, or harmful only in a certain environment; and the relevant environments are of two kinds: the external or secular, and the internal or genetic environment. What is good in the Arctic is not necessarily good on the equator; what was good in man in the ice age is not necessarily good now; what is good in a democracy is not necessarily good under a dictatorship. This is really so elementary for a geneticist that it would seem a waste of time to talk about it, if it were not for the fact that it is so often forgotten in discussions of human evolution, even by some geneticists.

The importance of the genetic environment is a little more subtle. In our laboratory shorthand, we talk about this gene being useful and favored by natural selection, and that one as being deleterious and discriminated against. Now, what survives or dies, produces offspring or remains sterile, is obviously not a gene but an individual with a certain genotype. It is the development pattern engendered by all the genes an individual has in a certain succession of external environments which confers a high or low fitness or causes lethality. There is no question but that some mutants give variants in man, as in *Drosophila*, which are unconditionally harmful, in all genetic and all external environments. I cannot think of an environment in which, say, retinoblastoma or hemophilia would be useful or even neutral. But even with apparently hopelessly harmful genes, surprises are in store. Professor Crow has mentioned the sickle-cell gene, which gives a lethal anemia when homozygous but confers protection against malaria when heterozygous.

We should not, however, confine our attention to the spectacular effects of mutations in the genes which are easily studied, just because their effects are so spectacular. The numerous variant genes with relatively small effects are probably more rather than less important in human populations. What proportion of them are unconditionally harmful and what proportion are harmful in some, but neutral or useful in other genetic and secular environments is a point on which we have little reliable information and which can be settled only by future studies. This is obviously a matter of fundamental importance for any evaluation not only of man's evolutionary perspectives but of the genetic structure of the existing human populations as well.

As it is, a geneticist resorts to a mathematical abstraction. We say that the retention or elimination of a gene variant depends on its average effects over the entire range of genetic and secular environments in which it occurs in a given population. This is a justifiable procedure for theoretical discussions of evolutionary genetics. I would like to suggest, however, that it does not tell us the whole story, especially where human populations are concerned. Consider the genes that cause a person's being especially prone to epilepsy. They are certainly harmful genes. As population geneticists, we say these genes diminish the Darwinian fitness and are discriminated against by natural selection. But it is possible that the genius of Dostoevsky was in some way conditioned by his suffering from epilepsy. Shall we deny that in Dostoevsky's genotype the gene that made him liable to epilepsy was perhaps of some value to the rest of mankind? We should not glibly assume that a genetic variant which is both biologically and sociologically detrimental in one genetical and environmental constellation will be so in all of them. We know far too little about human biology to equate the genetic leads with the sociological burdens of human populations.

J. F. Crow: Of course we need more knowledge. But we know much more about human genetics than animal breeders knew about their livestock two hundred years ago, when they were making such striking changes by selection. The principle that like begets like is much more often right than wrong. A program of selection may have surprises, particularly as to the *rate* of progress, but the general direction is usually quite predictable.

Garrett Hardin: I should like to call to your attention the large number of schizophrenics in the United States. If it is an overt 1 percent, this is a lot; and "overt" suggests that there are many other people whose personality needs looking into. Let me recall the situation with the sickle-cell trait. When it was first discovered, there was an attempt to explain the great incidence as a consequence of a high mutation rate, but we soon realized that this was ridiculous. Then Allison and others showed that we were faced with a polymorphism in which the heterozygote could survive but both homozygous types were eliminated—the one by anemia, the other by malaria.

Whenever we find a high percentage of an obviously disadvantageous trait—and 1 percent is high—we should wonder whether something of this sort is not involved. The heterozygote may be superior; or a genetic trait may have an advantage in some situations and a disadvantage in others. It is more than merely conceivable that a person with "schizophrenic genes" may not be an overt schizophrenic, but rather may become president of a corporation, a college professor, or a mathematician.

Without arguing about the evidence, we can agree that whenever we find a large percentage of some undesirable type, we should at least suspect that the genes involved may have an advantage under some circumstances. If we propose to get rid of these, we must first find out a great deal about the interactions of the genes and the various environments, as we have for the sickle-cell situation. We may find that we cannot get rid of a particular "bad" gene unless we literally modify society—just as before getting rid of the sickle-cell gene in East Africa we must get rid of malaria. Genes for schizophrenia probably pay off in some sense. We can probably find out the sense. When we do, we shall have to ask ourselves whether the "advantages" of the schizophrenia genes are so dear to us that we are willing to pay the price of some overt schizophrenia. Not until we face this question are we ready to act rationally. With increasing knowledge, I think we will discover many dilemmas of this sort.

H. J. Muller: If schizophrenia is due to characteristics which have been kept up by selection because they are advantageous and which we also value *per se,* if we tried to eliminate schizophrenia while at the same time preserving these other things, there would then be not so much to be afraid of, because, although trying to eliminate the schizophrenics, we would at the same time be selecting those very qualities which, when they come together in given circumstances, do give schizophrenia—not that we would want schizophrenia, but only those other characteristics. Thus, if that were the only way to have those other features, then we would continue to have schizophrenia, too, even though trying to eliminate it. In that case, we might in time decide that it was worth the cost, and not try to eliminate it genetically. As a matter of fact, I do not think that this representation of the situation is likely to hold true in the long run, because there are usually multiple pathways for getting a given phenotypic result. Of course all of us are *for* the individual. How could there be a good whole without good individuals? Yet it should be acknowledged equally that the individual is nothing without the whole, especially in the human species, and the sooner we all learn that, the better.

I. M. Lerner: Much has been made of the alleged success of early animal breeders in improving their population. This, I think, is a question that should be examined. They did not make any improvement until they had clearly defined the objectives. They knew exactly the characters they wanted. They did not talk about the ill-defined traits which are so subject to misinterpretation or, let us say, to different interpretations; they wanted a long body or a short tail or a pointed nose. This was an absolutely clear-cut goal, and they succeeded in attaining it only when the heritability of the desired characters was very high. It so happens that these morphological traits, traits of external appearance, do have a high heritability. The breeders also succeeded because they used exceedingly high levels of inbreeding—the basis of breed fixation.

On the other hand, there was a notable lack of success in such characters, as, for instance, the number of eggs laid by chickens, until the introduction of the more sophisticated methods of progeny and family selection. Even so, all kinds of complications and difficulties have beset breeders who tried to improve production. For example, I can give an illustration of the type of gene interaction, which apparently does not behave the way Dr. Muller and Dr. Crow tell us genes normally behave, that is, their effects change their direction under different conditions.

If you take an unimproved variety of chickens, you can select for the number of eggs and for the size of eggs and be successful in both. You are probably selecting for some aspect of metabolic capacity. However, if you start from, or reach a point of, high egg number, then selection for egg size may decrease egg production, which it did not interfere with in the first instance. There appears to be a conflict between the capacities to produce a large number of eggs and large eggs. So in one genetic background the two types of genes, or the two sets of genes that affect these properties work in the same direction; in another genetic situation they do not.

I think that this type of thing is absolutely bound to happen where you have selection for multiple objectives. If you have selection for a single objective, this kind of gene interaction, as you were told, very likely does not have any importance; but whenever you have selection for multiple objectives (and I do not see how in this particular discussion we can get away from the fact that we are to deal, not just with a single genetic character but with a combination of genetic characters which have sometimes negative correlations with one another) the much more difficult problem of interaction arises. I do not say that the goals involved are impossible, but I think we should question the optimism of obtaining results in a reasonably efficient and cheap way. Again, in breeding for such characters as egg number or milk production, the techniques used are progeny testing and sib testing. These are very expensive techniques in terms of genetic extinction, that is, in terms of preventing the reproduction of huge numbers of individuals in order to improve the trait by one or two percent.

Other complications arise when criteria of selection besides fitness are brought in. One of them is loss of fitness, which seems to be, at least in animal experimentation, a very common consequence of the selection of some metric objectives, even of high heritability. There is also very good experimental evidence for situations in which, given the same levels of inbreeding, fitness may or may not be lost, depending on the kind of selection used for traits other than fitness itself. There are also problems of correlated responses, one of the features of polygenic inheritance that has some bearing on this aspect. Darwin recognized this fact himself; he insisted on the existence of correlated responses, though what the mechanisms for it might be, and what if anything could be done about it, was something Darwin did not know. I do not disagree with the fact that it would be desirable to have a voluntary program for the elimination of genetic defects, or that it would be desirable to think or do something about improving certain qualities which we cannot perhaps define as yet and about which we have only some vague notions. All I intend is to point out that there are some technical complications of a genetic kind that we should not lose sight of. I think we would be doing a disservice to the whole cause, if we just glossed over them.

R. W. Gerard: I find myself in disagreement with something Dr. Crow said, that he did not believe there had been any strong natural selection toward intelligence or altruism. I would have thought that the evidence that intelligence, in the sense of richness of behavior, perception and modifiability, complex foresight and such, has a survival value is fairly clear.

Less clearly evident is an equally strong evolutionary trail of the survival value, in the strict Darwinian sense, of cooperativeness, togetherness, which at the higher level we call altruism. This is too long a topic to develop here; but the simplest quantitative thought shows that, unless some cohesive strength is greater than the dispersive strength, unless centrifugal forces are greater than centripetal ones, a system cannot exist and maintain itself in time as a coordinated group of units. This is true for a molecule, it is true for a multicellular organism, it is true for social groups at any level, and it is certainly true for human society. Once anything has appeared in evolution, making toward a greater interaction of parts, greater cooperativeness, greater altruism at the social level, it has had a demonstrable power of survival.

Ernst Mayr: In human evolution we must make a distinction between the early and the fairly recent evolutionary rates. We know that in the late Villafranchian era, when Australopithecus lived, hominid brain size was still around 650 cc., while some three or four hundred thousand years later, brain size had increased to 1500 cc. This was an extraordinarily rapid rate of evolution. But then the curve suddenly flattened out, and in the last 50,000 to 100,000 years there has been no increase in brain size at all. To the evolutionist, these facts suggest drastically different selection pressures before and after reaching the stable plateau. What was the selection force that resulted in the unprecedented increase in brain capacity? If one were a Utopian, one might ask why could we not restore the original rate of increase? I think there is no question whatsoever that when there were smaller human groups, the selective premium on altruistic traits and cooperative traits that helped the survival of a well-integrated group must have been exceedingly high. Today, however, in a big metropolitan civilization, even highly antisocial behavior is not especially severely punished by natural selection. So, I think, it is dangerous to generalize from one period in human evolution to another.

COMMENTS

G. G. Simpson: One way to judge in what direction evolution is now likely to be going under the influence of selection is to see what it has done in the past. This is a rather obvious approach. A lot of very poor popular science has been based on extrapolation from past experience. When we look back at the evolution of man, specifically, we do see that man arose by certain trends that can be identified quite easily in morphological terms. We know that morphological change was going on rather rapidly about twenty or thirty thousand generations ago. The rate of human evolution then seems to have been exceptionally high in comparison with the average for animal rates in general. The human rate then tended to slow down as time passed. From the skeletal evidence, and to some extent from behavioral evidence, it appears that at a relatively recent time, perhaps one or two thousand generations ago, the trends which had produced *Homo sapiens* thereafter slowed down, perhaps to an absolute stop.

From the point of view of evolution in general, as we see in many other kinds of animals, there is nothing extraordinary in this—in fact, it is the usual thing. We do have trends arising, usually moving rapidly at first, and then slowing down and giving way to stabilizing selection. Or, something may happen that turns an evolutionary trend to a different direction. In other words, the rate and direction of evolutionary change are not orthogenetic but are constantly changing under the varying conditions of selection pressures.

One can guess that the humanizing trends may possibly have stopped because they had exhausted the genetic possibilities. That would tie in with Crow's fascinating suggestion that there are impediments to artificial selection if its trend is in the same direction as that in which natural selection has already had time to act to the full. I doubt, however, whether this applies to the case of man. I suspect that if brain expansion stopped, as it seems to have done quite a long time ago, thirty thousand years or possibly more, it was not because there was no longer any genetic potentiality for further increase but because brain size had reached an optimum under the given conditions. Possibly this trend could be started again, but in any event it apparently is not going on now. If we decide that we want to enlarge human brain size, we cannot just say, "Well, that's the trend in man, and let her rip. Natural selection will take care of it."

Genetic Evolution

The apparent fact that stabilization in this respect occurred so long ago, at a time when human culture was certainly very different from what it is now and was by definition more primitive, suggests that we can almost assume that the stabilization that occurred then is not appropriate for the present time.

All this, of course, bears on the question of whether we really want to do something and should do something about human genetic evolution. My own feeling is that we should, because, if we do not, we probably are not going to like the results. That leads to some consideration of Muller's suggestions. For one thing, much of Muller's approach (not all) was based on the somewhat offhand assumption that we should eliminate all "defective" genes. That means, of course, we must further assure that we know what a defective gene is, that there is some absolute definition of defective genes. Perhaps there is. Perhaps Muller knows what a defective gene is and can clearly distinguish which genes are defective and which are not. In some cases, the distinction is certainly obvious, and no one could argue with him. In other cases, however, one cannot help wondering whether there is this nice dichotomy between defective and nondefective genes.

From the point of view, not of the individual, but of the population, would a genetically ideal population consist of individuals each of whom was genetically perfect? That is the kind of question I want to submit to the geneticists, who are eminently qualified to deal with it. In other words, the suggestion is that there is no one best genotype or even a few best genotypes, and that a gene that might in itself seem defective may nevertheless be desirable in the gene pool of the population as a whole. An established point that seems of first importance to me as an evolutionist who is not a geneticist is that natural selection is occurring in the human species at present. From the deficiency of data, doubts arise as to the direction and intensity of that selection.

A second point concerns the present direction of natural selection. We could probably agree that selection is doing some good as regards, for instance, the grosser genetical defects, which can be countered by natural selection supplemented now by available genetical counseling. In that regard, the present effect of selection is beneficial, but we cannot be confident that the total effect in other respects is good at the present time or will be good for the future of the human race. One does, of course, think that it would be nice if we could improve, genetically; this probably could be done only by some artificial selection, and thus one feels that some artificial selection in the human species would be desirable. The question then arises whether we know enough right now to warrant starting on this. My personal opinion, which is not authoritative, is that we do. Of course, we do not know everything, but it is a futile point of view to say that since we do not know everything we must not try to do anything. We do know enough to get started.

There arises then the very vital question (a political rather than a scientific one) of how to start. There is a great measure of agreement that the start must be made by those of us who see the possibility and the need. We must try to persuade people to make a start, voluntarily and individually. Even a small nonrandom element in reproduction can have important eventual effects; that is one of the principles of natural selection that fortunately can be transferred to artificial selection. A small measure of voluntary control of reproduction can increase positive selection for desirable traits and negative selection against undesirable ones.

Th. Dobzhansky: Being an incorrigible romantic, I am truly fascinated by the magnificent sweep and daring of Muller's imagination. Muller's utopia makes Huxley's *Brave New World* seem tame by comparison. This is, indeed, the bravest new world yet conceived, not only by a geneticist, but perhaps by any scientist.

I find myself in agreement with a part of Muller's arguments. It is indisputable that mutations do occur and that some of these mutations are unconditionally harmful, that is, detrimental to the organism in all the environments in which man lives or is likely to live, in combinations with all other genes present in human populations, and in homozygous as well as in heterozygous condition. Muller and Crow seem to believe that a majority of mutations are unconditionally detrimental. I am inclined to doubt this, and would prefer at least to regard this an open issue. But in any case, we agree that in human populations the accumulation of unconditionally harmful mutants is undesirable, that increases of mutation rates should if possible be avoided, and that even a total suppression of mutations (if this were possible, as at present it certainly is not) would be beneficial, at least in its short-term effects.

We also agree that the changes in the selective processes going on in human populations should be carefully watched. I am not sure that our evaluations of these changes are similar. It seems to me that the most we can wish natural selection to do for us is to keep us genetically adapted to our present environments, not to the environments of the Paleolithic or even of the Neolithic man. Our present environment is normal for us, and it includes such environmental components as clothing, housing, heating, modern technology, and modern medicine. The Paleolithic and Neolithic environments are now abnormal; we, and let us hope our descendants, will not live in these environments, and, if we are no longer fit to live in them, this may make us nostalgic but perhaps not unduly alarmed. Modern medicine saves some weaklings who would have been eliminated 100,000, 1,000, and even 100 years ago. But we may as well face it: we cannot and our descendants will not be able to dispense with the services of medicine and technology. Genetic changes induced by medicine will have to be taken care of by more medicine.

Of course, this does not mean that any genotype is as desirable in mankind as any other genotype, because some wonder drugs or wonder treatments will take care of everybody. It is all a question of the relative social costs of eliminating certain genotypes or of taking care of their carriers by special regimens. The lives of children afflicted with retinoblastoma can be saved by a special surgery, but it does not follow that it would be good for every child to be born with retinoblastoma and to undergo this surgery. I agree with H. J. Muller, therefore, that people who carry certain deleterious genes ought not to transmit these genes to posterity, and that we should hope that the spread of the genetic enlightenment will lead such people voluntarily to abstain, or at least to limit, their reproduction. (It is obviously impossible to draw a line separating those who should from those who should not reproduce.) This will depend on many things, of which not the least important will be the level of technology and of the medical arts reached in a given society at a given time.

Prophecies I wish to eschew, but it seems to me that people will find it much easier to agree on what genotypes are undesirable and should not be propagated than on what improvements should be striven for. The progress of negative eugenics is probably easier to achieve than an agreement on the qualities to be sought for in a genetically new man—despite the fact that the potentialities of positive eugenics (as Muller rightly points out) are tremendous and inspiring.

I fear that the ideas of the genetically new man very easily degenerate into something like the Platonic ideal Man with a capital M—something quite homozygous for the ideal genes and quite free from even a fraction of a lethal equivalent of the genetic load. It is too easy to let our imagination strive for something with a body as beautiful as a Greek god, healthy, and resistant to cold and to heat, to alcohol and to infections, with the brain of an Einstein and the ethical sensitivity of a Schweitzer, the musical talents of an Oistrakh and the poetical talents of a Shakespeare. In the first place, it is quite possible that all these qualities just cannot coexist in the same person, and that even several of them cannot, for genetic, or physiological, or psychological, or educational reasons. I do not mean that a beautiful body is necessarily incompatible with a beautiful mind, but rather that a peak performance in each field may depend on a very special genotype, and that some of these genotypes may be multiply heterozygous for numerous genes which produce rather less desirable, or even downright undesirable, effects when homozygous.

I believe Muller promises that his utopia will have some provisions for the maintenance of a variety of genotypes, so that life is properly spiced. If the history of science teaches us anything, it is the unwisdom of regarding a scientific goal as absolutely unattainable. Our remote descendants will perhaps reach so perfect a control of their genes that they will be able to produce genotypes with exactly the genes they want and to predict exactly into what sort of persons the carriers of these genotypes will develop. Until that goal is reached, I shall be inclined to believe that human populations should be genetically variable, so as to maintain their Darwinian fitness and also to produce a range of genotypes and phenotypes which is essential for the maintenance of human social life and creativity.

It seems to me that we should strive to achieve an equilibrium between two extremes: the one which believes that the genetic basis of society and of culture is either unimportant or will somehow take care of itself; and the second, which would presume our being in possession of a truly divine wisdom, sufficient to plan man's future from "here to eternity." Just where this point of equilibrium may lie is, of course, the great question.

R. S. Morison: In one sense, at least, I feel I have moved a little more in the direction of some of the things Professor Muller has talked about. I still feel that it is very hard to identify a bad gene. It is even harder to identify good ones, but nevertheless my attitude has changed a little, and for a somewhat paradoxical reason: in short, it is because I have confirmed my feeling that it may be harder to succeed than Dr. Muller suggested it was that I am happy to accept his conclusion that we ought to try.

As a side issue, there is a point I think no one has mentioned: what does it do to an individual who is asked to make choices in this area of behavior? In this connection, we may turn to the Greeks for advice, because it may prove that it will not be so much fun to try to choose between the good and the bad as we originally thought. For example, if the physicist may be thought to be suffering now from the guilt of Prometheus for giving man the gift of atomic fire, it is possible that those who apply or try to apply the laws of genetics may suffer from the same things as King Midas did; as he found out, when everything you touch turns to gold, you have nothing to eat. Similarly, it might turn out that parents who looked forward eagerly to having a Horowitz in the family would discover later that it was not so fine as they expected because he might have a temperament incompatible with that of a normal family. I think it is important, therefore, to think about the long-term effects of conscious genetic selection in a more serious way than is suggested by these joking references. Parents now often have a good deal of guilt because they have been told by the environmentalists that all the troubles of their children are owing to the way they have been brought up. This attitude has been summarized in a little epigram, that there are no bad children, only bad parents.

It is bad enough if we take responsibility only for the environment of our children; if we take responsibility for their genetic make-up, too, the guilt may become unbearable, and I do not think it is un-

COMMENTS

realistic to ask some of the humanists and philosophers and perhaps some of the psychotherapists to begin to think in terms of how we protect the next generation from this unbearable, or potentially unbearable, load of responsibility and guilt which will come when we can no longer put off on the shoulders of an all-wise Providence the defects which are inevitably bound to occur in our offspring and in ourselves.

D. H. Fleming: The main point I wish to make is that, if Dr. Muller's vision were to be acted upon, we would simultaneously have to alter most people's conception of the nature of modern science. It has been my observation that the great bulk of people, even the educated, think of science as infinitely permissive and almost immediately permissive in any problematical situation. We ourselves think of science as permissive also, but we take it for granted that, in the process of opening out more desirable prospects in the long run, it will commend and impose many restraints. Above all, we assume that the maximum benefits cannot be had from science without the continual application of foresight and forethought. But the permissive conception of science that I encounter among many educated people is entirely different: they appear to think we need not take any long-range thought for the future, because science will always stage an eleventh-hour rescue from any corner we may have boxed ourselves into. New supplies of food will be magically conjured up on demand, and new earths for habitation. It would not be exaggerating greatly to say that the triumphs of science, in this sense, have induced a general attitude of recklessness about the future of mankind. Men's age-old sense of the impossible, or at any rate of the dimension in time of the possible, has experienced a profound decay.

If, however, Dr. Muller's proposals were to be accepted, this would entail an imposition of restraints upon practical conduct in the name of science. People would be asked to surrender their pride in generating their own children and substitute for this pride the satisfactions to be derived from supplying a warm environment for children who are not biologically their own (at any rate, not the father's). Dr. Muller confidently expects the gradual emergence of a new superior form of pride and gratification in contributing to the

Genetic Evolution

genetic improvement of the human race. But we certainly have to anticipate that, in the early stages at least, an uneasiness at giving up one's claim to have actual biological descendants would be widespread, even among sophisticated people.

In this context, science would appear to the bulk of mankind for the first time as the bearer and commender of restraints upon practical conduct. I suppose one might think of the sanitary movement as having imposed restraints upon ordinary people from the early nineteenth century forward. Yet the sanitarians did not require of most people the surrender of anything they had previously regarded as particularly desirable. Filth and crowding had no positive emotional connotations. So it does seem to me that Dr. Muller's proposals would represent one of the first absolutely clear cases of telling people that they ought to abandon or restrict practical conduct of a kind that has had almost exclusively affirmative associations, and very strong ones at that. I do not bring this point forward as an argument against accepting Dr. Muller's proposals, but merely to emphasize that, in the degree that they are adopted, they will represent a passing over to science of the traditional role of religion as the fountainhead of restraints upon pleasurable conduct. I think that some such transfer of the burden of morality from religion to science is inevitable and desirable. We do, however, have to recognize that one of the tasks that would have to be performed in preparing people to accept Dr. Muller's philosophy would be to accustom them to the idea that science is not infinitely or cheaply permissive—that the long-term benefits it is capable of conferring must be paid for by accepting many present restraints and a perpetual re-evaluation of values in the light of scientific developments. It is precisely one of the incidental merits of Dr. Muller's vision, whatever one's ultimate judgment upon it may be, to stimulate thought upon this fundamental issue of the inexorable transfer of functions from religion to science. We may be certain that we are only at the beginning of the process.

Ernst Mayr: We want to clarify and simplify the concept of evolution. In principle, and according to the modern synthetic evolutionary theory, evolution is a two-step phenomenon: first, the production and maintenance of variation; and second, selection. This means that different genetic variants do not have the same probability of maintaining their frequencies from one generation to the next.

In the present discussion, variation has not been dealt with in detail, because everyone agrees that there is almost unlimited genetic variability in the human species, and that the only part of the variation we need to be concerned with, and possibly to worry about, is the deleterious variation. This deleterious variation, therefore, for the present at least, is that part of variation most important for the second step, selection.

With respect to this second part, I think that some of the participants, as I gather, were quite startled, if not shocked, by the novelty of some of the proposals and so-called Utopias. I think it has been wholesome and gratifying that the shock more or less wore off and that toward the end we have talked about things as though they were quite natural, though on the first day we considered them with a good deal of awe.

There are two aspects to this guidance. First, the negative selective aspect of the elimination of deleterious genes. At the present time this is the most feasible and the most necessary aspect, and it will presumably facilitate anything else that might come later. The other aspect is the possibility of positive selection, progressive selection, which for the time being should be strictly voluntary. There should be no interference whatsoever with human freedom. If the intellectual and moral climate should eventually change, there may no longer be a pure moral choice, but some sort of moral urging; this is something we will have to leave to the future.

Nobody could give a detailed blueprint of possible measures. I think we have become acutely aware that, much as we agree on the basic outline of the situation and on the need for action, our greatest weakness is the lack of many types of information that we would like to have. This concerns in part genetic information, and in part it concerns the interaction between the genotypes and the total environment, including the cultural environment leading to the particular phenotype which we consider human personality.

JULIAN H. STEWARD AND DEMITRI B. SHIMKIN

Some Mechanisms of Sociocultural Evolution

What is Cultural Evolution?

During the nineteenth century, cultural evolution provided the dominant frame of reference used to interpret the data of the social sciences.[1] Eminent scholars of the era, such as E. B. Tylor (1871, 1881), L. H. Morgan (1877), Friedrich Engels (1884), Herbert Spencer (1876-1880) and Leon Metchnikoff (1889) accepted the unity of the human mind and the common direction of human progress as basic maxims.[2] Without denying the special effects of environmental differences and the role of cultural diffusion, they stressed the central importance of universal evolutionary trends and stages. By placing particular societies in their appropriate stages, with due allowances for the effects of anachronisms or "survivals," both contemporary functioning and future developments might be predicted.

These theories have persisted in Marxist scholarship, and have been espoused by Leslie A. White (1949, 1959a, 1959b) and V. Gordon Childe (1951). But most anthropologists have followed the stern empiricism of Franz Boas and Bronislaw Malinowski (Lowie, 1937; Firth, 1957). The extensive ethnological fieldwork of these men and their students shattered early evolutionism by demonstrating the variety and uniqueness of innumerable cultural configurations and sequences, the diversity and relativism of cultural goals and logics, and the functional coherence of cultures in their own frames of reference. Even seeming anachronisms proved to have immediate significance in material or psychological terms.

JULIAN H. STEWARD AND DEMITRI B. SHIMKIN

The foundations of neo-evolutionism were laid in the 1930's and 1940's by the discovery of recurrent similarities in the social structures of simple hunting and fishing peoples; by the archeological demonstration of extensive parallelisms in the development of agriculturally based, urbanized states in the Old and New Worlds; and by the observation of independently replicated behavior, such as the rise of messianic cults, in response to similar types of socio-psychological stress. Cultural evolution has again become respectable (Steward, 1960); and intensive studies in this area have been undertaken. These have led to useful collections of essays celebrating the Darwin centennial—*Behavior and Evolution* (1958); *Evolution and Anthropology: A Centennial Appraisal* (1959); *Evolution After Darwin* (1960), and *Evolution and Culture* (1960), as well as many books and articles which are cited subsequently.[3]

These new studies differ from nineteenth-century views in several respects. Based on coherent assemblies of cultural data rather than the comparisons of isolated facts, they are more concerned with the growth or transformations of systems than with the origins of particular elements. Empirically oriented, they seek principles that are specifically applicable rather than universal, that is, sets of forces and patterns of interaction precipitating particular events rather than categorical causal sequences. Dynamic in purpose, they are focused more on recurrent processes than on broad historical reconstructions and taxonomies. While modern evolutionists accept the generality of certain trends, such as critical increases in economic productivity, they stress the multiplicity of structures manifest within these trends, in other words, multilinear evolution (Steward, 1953, 1955, 1960). Otherwise, they have reached little consensus on fundamental postulates concerning significant developmental criteria and processes. They now recognize that cultural conditioning joins with more constant biological needs, capacities, and limitations to produce human behavior. Attention to environmental limitations and conditioning, particularly the limitations on societies with simple technologies, is also widespread. In general, the new, empirical, and inductive approach might perhaps be designated as the "small or substantive view" of cultural evolution, as distinguished from the "grand view" of a century ago.

In this paper we shall attempt to conceptualize sociocultural

Sociocultural Evolution

evolution so that it has a maximum relevance to the practical problems of the contemporary and future world. Returning to a neoclassical position, we recognize the persistence in human history of several developments. Among these are expansions in productive capacity, leading to the growth of population. There is also an increasing dependence in productive activities upon cooperation and on the use of accumulated resources and knowledge (feed-back). Larger interacting communities with complex structures unify increasingly varied and specialized sub-units, while social controls change from categorical and direct means to conditional, flexible, and symbolic ones.

Yet we also stress the spasmodic, discontinuous nature of shorter-run events, the indeterminacy of specific developments, and the inconstancy of other cultural trends, particularly in human values and modes of expression. We shall specify our working hypotheses on the mechanisms of sociocultural evolution by successive statements on our general propositions, on the flow and organization of culture, on cultural institutions as dynamic factors, on some types and levels of sociocultural integration, and on current problems of prediction and control. But we plead that the empirical researches and conceptual critiques needed to develop and evaluate a genuine understanding of cultural evolution have barely begun.

Some General Propositions

Our treatment of culture and its evolution rests upon nine heuristic concepts which constitute a mixture of hypothetical postulates and real but tentative observations. In essence, they cover (1) the basic components of culture, (2) the origination of cultural elements, (3) the nature of culture patterns, (4) the composite character of whole cultures, (5) levels of sociocultural integration, (6) cultural ecology, (7) interrelations between culture and biology, (8) process and culmination, and (9) cultural indeterminism.

1. Culture summates the body of adaptive, cognitive, and expressive behavior and its consequences, developed and socially transmitted by man through space and time. It consists of a flow of individual elements subject to high rates of innovation, change and disappearance, and of more stable relational systems of operational

patterns which sift, redirect, and modify this flow toward more coherent and purposive efforts. The central problems of cultural evolution are, accordingly, dual: the life histories of elements, and the identifications of patterns, including their development and transformations.

2. We believe that cultural innovations are constantly developing as results of transmittal errors, intellectual play with existing patterns, contextual changes and deliberate disjunctions. However, the acceptance of innovations is rare, since it depends upon the social recognition of their distinctiveness, their utility (physical or psychological), and their compatibilty with existing practices. The other source of culture (borrowing) has effects which also are greatly conditioned by the nature of the transmission, the elements borrowed, the prior presence of competing elements, etc. Moreover, borrowing may take place in symbolic form through stimulus diffusion, or physically diffused elements may acquire novel functions in new environments. For these reasons, analysis of any culture history always retains much uncertainty.

3. Cultural patterning reflects basic operating systems in human neurophysiology (von Neumann, 1958; Brazier, 1959). Especially important is the phenomenon of perceptual facilitation, by which stimuli of habitual significance are better discriminated than others having equal or greater physical strength but lacking in subjective meaning (Rosenzweig and Postman, 1958). In fact, among the Navaho and Hopi, linguistic conditioning appears to have led to a dominance of an entire perceptive mode, namely, form, over discrimination as to color and size (Carroll and Casagrande, 1958). Patterning is also determined by physical requirements, e.g., metallurgy presupposes mining or trade.

The span and cohesiveness of patterning vary among cultures; patterns are ubiquitous in kinship structuring, oral literature, semantic associations, and similar bodies of relatively homogeneous elements. Characteristically, patterns combine microvariations (individual styles) with prescribed rules (Shimkin, 1947). Above all, despite such resistance to change as is offered by the normative aspects of integrated patterns and the culturally conditioned human personality, individuals do change cultures within a lifetime, and societies may change within a generation or two, as in many contemporary in-

Sociocultural Evolution

stances of tribal societies which have come under the influence of the industrial world. Culture, then, is phenotypical behavior. While biological analogies may be stimulating, it is not expectable that the principles of biological evolution can be applied to cultural evolution.

4. Certainly, over time and space, no culture is a closed unit, for the dynamics of technological, linguistic, and social evolution disclose different processes, rates, and structures; moreover, there are always some conflicts and imbalances within every system. Thus, considerable latitude is found for selecting significant or diagnostic features, so that evolutionary taxonomies have been variously based upon technology, economics, kinship systems (especially by British writers), political patterns, and even value systems. We feel that all these institutions must be disaggregated from each culture and studied as homogeneous groups. At the same time, quite general, composite criteria may be set up for broad temporal-functional groupings, which we call levels, as distinguished from types of sociocultural integration.

5. The concept of sociocultural levels is a heuristic means of analyzing developmental sequences and internal structures. The approach seeks hierarchies both of organization and of sequence—the autonomous nuclear family, extended families of various types superimposed upon the nuclear family, or a multicommunity state unifying hitherto independent settlements. Each level generates new properties and principles of operation, which govern the new assembly and modify the older component units. Thus, the roles and properties of nuclear families of similar structure are profoundly different among the Eskimo, where the institution is essentially independent, and in the modern United States. There, the family operates in a vast complex of legal and political constraints. We suggest that these levels are structural and functional; hence, they may occur in groups with entirely different histories and differentiate others of single origin. In so far as the concept of levels of sociocultural integration represents real groupings, these are basically convergences toward similar solutions of like ecological problems.

6. Cultural ecology signifies the social adaptations entailed when a given technological inventory is used for exploiting the resources of a specific environment. Such frameworks set the basis for cooperation, aggression, and other forms of social in-

teraction. The adaptations are most direct among the simpler, hunting-and-gathering societies, where social structure must respond in high degree to exigencies of the quest for food. The nuclear family may cooperate principally with collateral relatives on both sides, as among desert seed gatherers; with patrilineal kinsmen, as among many hunters; or with matrilineal kinsmen, as among many simple farming societies. Among more complex groups, the sociocultural environment sets the major contexts, which are of a different nature than the physical environment. Thus, the variety of possible social arrangements tends to increase, while the requirements of land use (for mines, farms, transportation, cities, etc.) become more rigid.

7. Implicit in the concept of cultural ecology is a recognition that human biological characteristics underlie culture. Prominent among these are sexual reproduction, prolonged infancy, a high capacity for learning, including a marked ability for symbolization, extreme climatic adaptability, omnivorousness and a capacity to withstand starvation, and probably some aspects of emotional organization. Sex, age, and kinship strongly obtrude in the organization of the subsistence-oriented egalitarian societies which were ubiquitous before the rise of agriculture.

At the same time, the relations between biology and culture are extremely complex. There are no grounds for postulating that forms of cooperation or aggression are determined by the genetic character of the group. Primitive people characteristically cooperate with a locally defined in-group, such as the lineage, band, or community, and tend to display hostility toward most outsiders. Recently, individuals in many such societies, have come into competition with one another owing to the introduction of a cash economy and the shortage of lands, but they have begun to cooperate economically and socially with the outside world. In addition, most societies have some means of inhibiting deviant individuals, and of facilitating favored behavior (Thurnwald, 1931-1935, vol. 4, pp. 264 ff.). The relations between inherent temperament and *pro forma* behavior are therefore difficult to disentangle; latent characteristics, such as undiscovered genius, may be fully hidden.[4]

In general, human variability seems to have had no proven effect on culture. The basic elements of culture—social groups and definable traditions of toolmaking which necessitated the use of language—were produced by proto-men, perhaps Australopithecines. Among living men, efforts since the days of phrenology have produced no statistically demonstrable physical or physiological correlates (pathology apart) of intelligence, specific aptitude, or temperament. Nor have any correlations been advanced between these features and the many types of breeding groups in man. The accidents of population contact or isolation, esthetic preferences, social prejudices, and taboos have unquestionably influenced man's genetic composition. Yet culture, so far as all evidence shows, diffuses and flowers under equivalent conditions among all groups.[5]

8. Cultural processes take place at times and in regions that are defined by rates of pattern development, element innovation, and diffusion. In ideal cases, patterns evolve from accidental associations into tested, rationalized codes applied to a variety of circumstances, and then degenerate into stereotypy. The maximum of patterned standardization, organization, and productivity, we term culmination. A full-fledged absolutistic empire, predatory band, or other social type may have diverged considerably from its incipient stages, yet it represents the culmination of the same processes. This fact underlines the importance of identifying processes and perhaps makes them basic in typology. Another aspect of the time-space relations of cultural evolution is the great importance of relative sequence, which may lead to radically different consequences. For example, the exposure of Japan, with its well-developed foundation of elemental industrialization, to complex industrialization was followed by rapid selective borrowing and an adaptive restructuring of existing social and economic institutions (Smith, 1959). In most of Africa, without these foundations, complex industrialization cannot be readily assimilated.

9. Finally, cultural evolution is significantly indeterminate, so that reliable reconstructions of events, let alone predictions and programs, can be achieved only to limited degrees. Three causes contribute to this indeterminacy. One is the complex nature of cultural events, which, as we have indicated, reflect a variety of influences of the present and past. The second is the fact that most cultural phenomena may be generated in more than one way; for example, the number 8 may represent $4 + 4$, 2^3, or $10 - 2$. The

third is that transmission errors and information losses, sequential changes, and other factors make cultural forces and patterns imprecise; their products and derivatives become increasingly so.

Under these circumstances the relative regularity of culture is remarkable. It is accountable in part by the requirements of underlying biological invariants, but even more by the normalizing powers of learned patterns, especially when reinforced by stored materials and systematic training. Thus, the best guide to cultural prediction and guidance rests in the painstaking analysis of these patterns, their contexts and their interrelations. Even then, great caution is essential.

The Flow and Organization of Culture

The mechanisms of cultural evolution may first be viewed with reference to culture elements. These consist of developments within communities and of transfers and interactions between communities. The former comprise the life histories of cultural elements, including innovation, acceptance, transmission, and loss; also, the dynamics of patterns. The latter consist of diffusion and acculturation.

Intensive investigations in many cultures show that innovations are frequent everywhere, but that the types and degrees of innovation which may be favored or inhibited vary greatly (Barnett, 1953). Among the Shoshone, for example, novel, supernaturally sanctioned distinctions in ritual are valued, as were original poems in medieval Europe or Japan, and as scientific creativity tends to be among us today. In general, institutionalization appears to be most effective in stimulating innovation, although intellectual play, emotional satisfaction, and pragmatic utility are also influential. Acute physical or cultural deprivation rarely generates effective responses; instead, flights from reality as exemplified by messianic cults are more common. At a later stage, survivals of elements of the crisis period may be integrated into new cultural complexes (Shimkin, 1942).

The acceptance of innovations begins with a positive evaluation which results either from a rational discrimination of a perceived or supposed advantage in the new item or type of behavior, or from its association with the prestige of a person, a group, or an event.

Sociocultural Evolution

Acceptance is achieved when the innovation has been symbolized, replicated, and transmitted to others. In all societies, demonstration and the didactic process promote the transmission of selected elements of culture to the young. Teaching is often facilitated by supernatural, physical, or social sanctions against error, as well as by structured contexts such as isolated schools or communication via folklore (Pettit, 1946). Above all, the perpetuation of innovations is greatly influenced not only by their intrinsic merits and compatibilities with the culture, but also by their relationship to the status, communication, and educational systems of the society.

Cultural attrition, complete loss, and replacement have multiple causes. In small, nonliterate communities which lack special mnemonic devices, such as genealogical records, dispersions of population may lead to greatly accelerated rates of transmission error and loss. Associated with dispersion (and sometimes with climatic changes) are re-adaptations. For example, the Samoyedic peoples, who were horse-breeders in the first millennium B.C., borrowed the practice of reindeer riding from Altaic peoples in the course of their northward migrations. On reaching the tundra, they found that the northern reindeer were too weak to ride but were more effective for pulling sleds than the aborigines' dogs. Finally, when they had killed off most of the wild reindeer, the Samoyeds expanded their herds to serve as major sources of food as well as means of transportation (Shimkin, 1960). In similar fashion, the Micronesians, when no longer able to procure tough stones for tools and weapons, substituted giant clams and sharks' teeth. Even without dispersal, most societies are quick to seize opportunities for rational substitution, for example, they use galvanized iron for roofing in place of labor-consuming, vermin-infested thatch. The active suppression of emotionally weighted elements, such as the abolition of the fez and the veil in Kemalist Turkey, is common in situations of crisis or conquest. Conversely, unsuccessful repressions intensify the durability of symbolic elements, which may be revived during a period of secondary reconstruction, as was the Confederate flag. Finally, the loss of a key element often precipitates the destruction of an entire associated complex.

The persistence of cultural elements is strengthened by integration into patterns, which always have two aspects of rather different composition. One is a rather undifferentiated, nonrationalized, psychological image (the cultural complex) in which specific visible elements (cues) are highly transmissible and diffusible, but in which the functions and internal organization are unstable. The sun dance of the North American plains and the naive image of recent Western progress, which we call the Victorian complex, typify this class. The other is the rationalized pattern, the core of which is a defined operational code applied to varying contents. For example, the pattern of a commercial corporation is that of a legally specified code which is applied to a wide variety of specific contents. In technology particularly, rationalized patterns conform closely to physical requirements; in other spheres, such as kinship, logico-esthetic considerations such as symmetry, comprehensiveness, and closure appear to be paramount.

Patterns may also be distinguished by their characteristic histories, which are largely correlated with particular aspects of culture. First, cumulations are characteristic of the technology of production, since more advanced applications presuppose earlier knowledge. The curve of technological development is not smooth. There are bursts and periods of stagnation, but losses or radical replacements are rare. Technological progress tends to be replicated many times, just as it is readily diffused, because it meets basic needs. Also, it tends to carry similar concomitants, such as population growth and nucleation, social specialization, and the development of political controls, religious sanctions, and communications.

Second, "cyclicity" governs a large category of elements, especially those which are biocultural and expressive, such as etiquette, gestures, humor, recreational patterns, and many others not essential to technology, which seem to be accessory, rather than organically related to social types, and which often have many possible forms. These, it seems, can readily be added to, subtracted from, or altered within the total culture. Perhaps art styles and elements should be included in this category, since they seem to evolve through pulses, and they may readily diffuse, without substantial alteration of other features. Art pulses differ from those of technology, since they do not necessarily have any continuity with the past.

Third, the techniques and uses of communications, like the technology of production, are cumulative, but their evolution is marked by a few sharp changes, such as the introduction of writing and the establishment of libraries. While changes in the features of the second category do not seem to entail evolution in other categories, the evolution of communications has been a critical factor in the changes in many major aspects of culture. At the same time, the permanence of the recorded word tends to be a stabilizing or normative factor.

Fourth, and most important for present considerations, are transformations in social structure, which we define as the ways in which individuals relate to one another, both in the context of the family and in relation to organizations of greater magnitude, to maintain subsistence and achieve economic, religious, political, recreational, and other social goals. Social transformations represent the emergence of qualitatively new levels of sociocultural integration more importantly than do the accretions or accumulations of elements, as in the previous categories. These emergences are brought about by the processes already discussed and by the results of social contact.

The latter term comprehends an immense range of situations, from the transmittal of an isolated element to the physical amalgamation of populations; it may be further complicated by friendliness or hostility, patterns of equality, or dominance-submission relationships, the presence or absence of a common language, etc. The consequences of contact may be generally classified as diffusion and acculturation. Diffusion is related to the introduction and the modification of received elements or patterns; these are processes which are manifested in many forms and degrees, almost always selectively. Acculturation concerns the transformations brought about in receiving societies by diffused elements or patterns, or secondarily induced by various conditions of contact. Typical acculturational phenomena brought about by the impact of the Victorian complex in Africa and Latin America include reduced mortality, physical and social mobility, individualization, and xenophobia.

In sum, the mechanics of cultural evolution are highly varied and complex, particularly so when they result from culture contact; however, they may be common to societies of various periods and levels of integration. Cultural emergents, or disjunctions in properties and processes, are most concentrated in the dynamics of specific cultural institutions.

Cultural Institutions as Dynamic Factors

Our studies to date indicate that cultural evolution has been particularly active among four types of institutions—subsistence activities, the control of energy and materials, the regulation and organization of populations, and communications—which together comprise the material and structural basis of society. Moreover, except in simpler societies and within limited areas of influence in more advanced ones, such biological principles as sex, age and kinship become secondary to status, occupational, role and other socially-derived principles.

Subsistence activities comprise food production, its handling, storage, and preparation, with auxiliary efforts, such as the care of draft animals. Although subsistence activities dominate the cultures of simple hunting, gathering, and fishing societies, they remain important in all others as well. Thus, slash-and-burn agriculture, animal-powered field cropping, and large-scale sheep-raising require distinctive patterns of land use, settlement, labor utilization, and annual or periodic cycles of activity. Moreover, complementary efficiencies must characterize the various phases of the quest for food to ensure stable supplies. For example, without a shift from querns to the more productive rotary or rocker mills, the increased yields from plowed fields can scarcely be utilized. In general, the gains in output and in labor productivity in subsistence activities provide the foundation for all other surpluses and for leisure.

The control of nonhuman energy and materials depends upon the making of tools, transportation equipment, measuring and control devices, and such items of direct use as shelter and clothing. While tool-using is the first measurable indication of culture (the control of fire is almost as old), development in the control of energy and materials has been spectacularly rapid since the Middle Ages. Today, an increasing mastery of energy makes possible unprecedented scales of production and destruction, new materials, and a society based on specialization rather than stratification. The development of modern materials extends man's motor and sensory limits far beyond his biological endowment, while freeing him increasingly from climatic and other environmental stresses.

Increasing supplies of food, energy, and materials have led to proportionately larger populations, since man has long limited his numbers by many devices wherever shortages in resources have been severe but persistent. Until recently, disease was a far more intractable problem, and even today the worldwide major gains have been realized only against infections. The total of the population, however, is less important than the size and nature of population aggregates, which must be explained by cultural-ecological adaptations and by the varied economic, political, and military functions of population centers. Transportation routes are similarly derivative.

The factors integrating the numbers and physical distributions of populations are the systems of communication, including direct interactions. With technological development, communications has successively included: standardized symbolization, as in language; memory, as in writing, photographs and records; increasingly full spatial transpositions of thoughts and influences, by messenger, letter, telegraph, telephone, radio and television; and protracted calculation, by computer. Advances in communications have been most important in generating new, often paradoxical properties in culture. In particular, mass communications, in which leaders and persuaders transpose their images into direct, simultaneous contact with enormous groups of viewers and listeners, has promoted both the high centralization of authority and individual judgment.

Warfare is another institution which evolves and changes its qualities in different contexts. Militarism for the conquest of territory and the control of people and wealth, though characteristic of recent millennia, was probably little known to primitive man, who engaged in blood feuds or in expeditions to capture sacrificial victims. Modern warfare, therefore, presents qualities not foreshadowed in early society. Its crucial aspect is the potential for total destruction, which evolves both from modern science (itself an emergent quality) and from its social context and purposes.

The nature and social functions of religion have also evolved. To the personalized relation between primitive man and supernatural forces (especially guardian spirits, the belief in good and bad natural forces, and the shamanistic function of treating sickness), there was added ceremonialism, for such group purposes as procuring food. Religion at higher levels of sociocultural integration served to sanction national social and political purposes. A still different kind of religious emergence seems to have been the personalized or messianic religions that began about 600 B.C.

These and other processes have evolved or acquired new qualities within their total contexts, while contributing to the evolution of this context. From a taxonomic and explanatory point of view, our difficulty at present is that comparative studies are too little advanced to permit an unequivocal assessment of such factors as productive surplus, types of settlements, communities and urban centers, militarism, religion, political systems, and others. Even the effort to construct a typology of modern nations is in its infancy.

Some Types and Levels of Sociocultural Integration

Early society, beginning with the palaeolithic period, constituted a broad category of kinship-based, subsistence-oriented communities. We have already indicated that the structural principles of such communities are strongly marked by the biological facts of sex, age and kinship. The nuclear family is structured on these principles, and serves the purposes of sex, procreation, the sexual division of labor, and the socialization of offspring, which have prolonged pre-adult growth. The particular extension of kinship relations to other relatives may be influenced by feasible groupings and adaptive requirements of specific subsistence methods in particular environments. Hence, it is possible to recognize such types as independent families of food gatherers (Shoshoni and Alacaluf), patrilineal bands or extended patrilineages of cooperative hunters of sparse, nonmigratory game, matrilineal bands, and probably other types (Steward, 1938, 1955b; Steward and Faron, 1959).

These primitive groups are essentially egalitarian in that each person behaves in approximately the same way as do other members of his sex, age, and kin group. With improved subsistence, there comes a point of surplus production—which however, has never been adequately defined—when specializations of role and differences in status begin to enter the society. These introduce entirely new principles of social structuring. While there are many instances of incipient breakthrough points, when egalitarian kinship and hereditary rank or status lineages come into conflict, no comprehensive study has isolated the common factors. In East and West Africa, it is possible that diffused patterns of kingship and militarism created royal lineages within clans which arrogated the slight surplus to themselves. In Polynesia, as Goldman (1955, 1957, 1958) suggests, a pattern of status rivalry was a precondition of status lineages, which arose where production was adequate. Certain lineages among the Chin of Burma, according to Lehman (MS), acquired surplus and special status through a marriage system which gave them control of lands, trade, and loot from warfare. On the northwest coast of America, warfare may have been a factor, but the status lineages also seem linked with an entrepreneurship which captured control of the fur trade.

At present, therefore, one cannot say with certainty that the breakthrough to hereditary status was achieved through a single line of evolution or that it was caused by any one factor. Population density per se and the presumed productive efficiency underlying a dense population do not necessarily lead to status lineages, as is exemplified in many areas. It is possible that diffused stimuli are needed to crystallize the new patterns. Lehman suggests, for example, that the Chin derived their status goals from the larger Burman civilization. But we are still left without an explanation of the first hereditary classes, occupational specialization, state political controls, and national religions which evolved in the great civilizations of antiquity.

These early civilizations were based upon a very efficient agriculture, much of it irrigation farming. There has been a controversy concerning the role of managerial controls of irrigation in state development (Steward, editor, 1955a). The factor of demography is related to efficient production. Increasing population may have stimulated the expansion of irrigation, as well as the reverse. But there are other factors which are not clearly understood. Changes in the nature of warfare and its role in state-building need more comparative study in the Andes, Meso-America, the Near East, and China. There also appear to have been important regional differences in the concepts of property, the control of production, trade, and markets, and in the introduction of money. There were obviously great differences in religious systems, despite many functional similarities in their relation to the state, especially in their possible nucleating and integrating functions.

While the great civilizations of antiquity were evolving, farming, technology, and also, perhaps, social patterns such as warfare and kingship, diffused from them and formed a basis for state development elsewhere. These have been described repeatedly, but no endeavor has been made to construct a comparative taxonomy. Reischauer (in Coulborn, 1956) has shown that feudalism in Japan and in northern Europe were extraordinarily similar, but were very different from that of India. In addition, many societies of mounted nomads also developed states, which reacted upon agrarian states in varied ways.

Our brief paper cannot detail culture history. We have found, however, that increased productive efficiency is a condition for the rise of larger societies. Other formants include military force, religious sanctions and controls, national symbolism, codification of national laws to supersede varied local customary usages in areas of national concern, formal state institutions; and generally, an invasion of and modification by national institutions at lower levels of sociocultural integration. Beyond this, we deal with a varied array of specific situations, institutions, and patterns. Among the most important of these has been industrialization (Shimkin, 1947).

Problems of Prediction and Direction

Today's conflicts and uncertainties reflect a concatenation of cultural-historical processes. Two world wars have redistributed global military and economic power, nullified prior conventions of international behavior, and upset domestic systems of status and authority. A new ideological force, namely, a utopian, elitist and militant, secular religion, controls one third of the world and influences the rest. Novel means of mass destruction and scientific subversion threaten every society. Unprecedented rates and confused sequences of change in developing countries overwhelm their adaptive capacities; the reduction of mortality without birth control and the appearance of physical mobility without new jobs illustrate this confusion. Independence brings new anticipations and tasks to nations lacking in needed means, institutions, or responsible leaders. Mass communications project the hostilities and anxieties thus generated, promoting even greater tensions. At the same time, new adaptive possibilities are emerging. Science and technology now can amass, disaggregate, restructure and apply vast quantities of information. Major expansions in the availabilities of food, energy, and materials have become possible through inventions and discoveries, such as hybridization, desalinization of water, etc. The major industrial nations now possess enormous reserves of manpower, capital and materials, which can be diverted to emerging needs without serious sociopolitical strain. And international instrumentalities, although still weak, provide, at least, face-to-face forums and specialized cooperative activities transcending national and ideological hostilities.

The future holds a spectrum of alternative possibilities, ranging all the way from human extinction to an integrated world community, with limited conflicts, continuing chaos, and regional stabilization as the more likely intermediate courses. Progress toward the ideal of a just and stable world society, an ideal which holds many unresolved contradictions, can at best be slow. In one aspect—the guidance of developing nations—past and present experience affords some help. Effective programs must modify cultural ecologies to meet urgent human, institutional, and technical deficiencies in ways least disturbing to existing structures. They must lead to new levels of sociocultural integration by redirecting aspirations toward larger joint tasks, the accomplishing of which would be facilitated by external resources, common institutions and new statuses. And they must ensure the culmination of these processes by proper civil and military safeguards against interference. Implicit in these suggestions is the need for great expansion in data-gathering, analysis, and trial in the decisive field of cultural behavior as applied to contemporary problems. But even with the best of knowledge, the inherent indeterminacies of culture demand that goals always be limited and programs be kept flexible.

In sum, the past offers little precedent for the solution of such urgent problems. At the same time, mass educational techniques and other mass media of communications afford the means of achieving change far more rapidly than before. A discriminating course of progress, which avoids both passive drift and reckless experimentation, is perhaps needed. Meanwhile, in this discussion, we have tried within our competence to formulate a means of identifying the determinants of sociocultural evolution in the past and of adumbrating some of the implications of such understandings. One important goal of research on cultural evolution will be the determination of how social values and objectives (including those of the leaders and men of good will who would guide our destinies) have evolved. What we have formulated is at best a beginning; we hope it will be at least a stimulus to constructive thought and effort on a vital problem.

REFERENCES

1 This version of our paper has benefited in concept and terminology from the discussion at the conference and from the editorial help of Betty W. Starr.
2 The dates are those of the first editions, those used here. On Marx and Engels, see also Bober (1950).
3 See Roe and Simpson, 1958; Meggers, 1959; Tax and Collender, 1960; and Sahlins and Service, 1960.
4 The relations between "cyclicities" in the frequency of recognized creativity and broader patterns of culture history in the Western world have been carefully studied by Kroeber (1944). They appear to be functions of pattern growth and culmination rather than independently appearing causative agents of change.
5 Any genetic determinants of human behavior, however great they may appear in certain individual differences, cannot at the present time be ascribed a causative role in cultural differences. It is entirely possible that societies may have genetic differences that are comparable to those of individuals, but such differences are so slight that their existence has never been established, while their role in determining cultural patterns is so infinitesimal that they can be disregarded. A demonstration that genetic factors have shaped cultural patterns will require a rigorous scientific methodology that has not yet been developed.

The assumption that individuals can be bred for a superior culture not only lacks scientific validation of the relation between genetics and culture but presupposes indefensible conclusions concerning the superiority of any culture. There are no ethical grounds for maintaining that modern culture is inherently superior to primitive culture or that either science or philosophy can blueprint a better culture for the future.

We therefore strongly disagree with the recommendations made by Professor Muller in his paper before Conference A.

BIBLIOGRAPHY

1. Homer G. Barnett, *Innovation: The Basis of Cultural Change*. New York: McGraw-Hill, 1953.
2. M. M. Bober, *Karl Marx's Interpretation of History*. 2nd edn., rev. Cambridge: Harvard University Press, 1950.
3. Robert J. Braidwood, "Archaeology and Evolutionary Theory," in Meggers, pp. 76-89.
4. Marie A. B. Brazier, *The Central Nervous System and Behavior*. New York: The Macy Foundation, 1959.
5. John B. Carroll and Joseph B. Casagrande, "The Function of Language Classifications in Behavior," in Eleanor E. Maccoby *et al.*, editors, *Readings in Social Psychology*. 3rd edn. New York: Henry Holt & Company, 1958.
6. V. Gordon Childe, *Social Evolution*. London-New York: Henry Schuman, 1951.
7. Rushton Coulborn, *Feudalism in History*. Princeton: Princeton University Press, 1956.
8. Friedrich Engels, *The Origin of the Family, Private Property and the State*. New York: International Publishers, 1942.
9. Lucienne Félix, *The Modern Aspect of Mathematics*. Translated by Julius H. Hlavaty and Francille H. Hlavaty. New York: Basic Books, 1960.
10. Raymond W. Firth, editor, *Man and Culture: An Evaluation of the Work of Bronislaw Malinowski*. London: Routledge and Kegan Paul, 1957.
11. Irving Goldman, "The Alkatcho Carrier: Historical Background of Crest Prerogatives," *American Anthropologist*, 1941, 43: 396-418.
12. ——— "Status Rivalry and Cultural Evolution in Polynesia," *American Anthropologist*, 1955, 57: 680-697.
13. ——— "Cultural Evolution in Polynesia: A Reply to Criticism," *Journal of the Polynesian Society*, 1957, 66: 156-164.
14. ——— "Social Stratification and Cultural Evolution in Polynesia," *Ethnohistory*, 1958, 5: 242-249.
15. W. H. Goodenough, "Oceania and the Problem of Controls in the Study of Cultural and Human Evolution," *Journal of the Polynesian Society*, 1957, 66: 146-153.
16. William A. Haag, "The Status of Evolutionary Theory in American Archaeology," in Meggers, pp. 90-105.
17. A. Irving Hallowell, "Behavioral Evolution and the Emergence of the Self," in Meggers, pp. 36-60.
18. N. B. Hawkins and C. S. Belshaw, "Cultural Evolution or Cultural Change—The Case for Polynesia," *Journal of the Polynesian Society*, 1957, 66: 18-35.
19. Julian S. Huxley, "Cultural Process and Evolution," in Roe and Simpson, pp. 437-455.
20. Melville Jacobs and Bernard Stern, *Outline of Anthropology*. College Outline Series. New York: Barnes and Noble, 1947.

21 Clyde Kluckhohn, "The Role of Evolutionary Thought in Anthropology," in Meggers, pp. 144-157.
22 Alfred L. Kroeber, *Configurations of Culture Growth*. Berkeley-Los Angeles: University of California Press, 1944.
23 ———— "History and Evolution," *Southwestern Journal of Anthropology*, 1946, 2: 1-15.
24 Eleanor Leacock, "Social Stratification and Evolutionary Theory: A Symposium. Introduction," *Ethnohistory*, 1958, 5: 193-199.
25 Anthony Leeds, "Architecture in the Evolution of Culture—An Anthropologist's View," *Balance*, 1957, 4: 3-5.
26 Frederic K. Lehman, The Chins of Burma (manuscript, n.d.).
27 Alexander Lesser, "Evolution in Social Anthropology," *Southwestern Journal of Anthropology*, 1952, 7: 134-146.
28 Robert H. Lowie, *The History of Ethnological Theory*. New York: Farrar and Rinehart, 1937.
29 Margaret Mead, "Introduction to Polynesia as a Laboratory for the Development of Models in the Study of Cultural Evolution," *Journal of the Polynesian Society*, 1957, 66: 145-146.
30 ———— "Cultural Determinants of Behavior," in Roe and Simpson, pp. 480-503.
31 Betty J. Meggers, editor, *Evolution and Anthropology: A Centennial Appraisal*. Washington: The Anthropological Society of Washington, 1959.
32 Léon Metchnikoff, *Civilisation et les Grandes Fleuves Historiques*. Paris: Hachette et Cie., 1889.
33 Lewis H. Morgan, *Ancient Society*. New York: Henry Holt and Company, 1877.
34 George P. Murdock, "Evolution in Social Organization," in Meggers, pp. 126-143.
35 George A. Pettit, "Primitive Education in North America," *University of California Publications in American Archaeology and Ethnology*, 1946, 43: no. 1.
36 M. R. Rosenzweig and L. Postman, "Frequency of Usage and the Perception of Words," *Science*, 1958, 127: 263-266.
37 Anne Roe and George G. Simpson, *Behavior and Evolution*. New Haven: Yale University Press, 1958.
38 Marshall D. Sahlins and Elman R. Service, editors, *Evolution and Culture*. Ann Arbor: University of Michigan Press, 1960.
39 Demitri B. Shimkin, "Dynamics of Recent Wind River Shoshone History," *American Anthropologist*, 1942, 44: 451-462.
40 ———— "Wind River Shoshone Literary Forms: An Introduction," *Journal of the Washington Academy of Science*, 1947, 37: 329-352.
41 ———— "Industrialization: A Challenging Problem for Cultural Anthropology," *Southwestern Journal of Anthropology*, 1952, 8: 84-91.
42 ———— "Western Siberian Archeology: An Interpretative Summary," in Wallace, pp. 648-661.

Sociocultural Evolution

43 Thomas C. Smith, *The Agrarian Origins of Modern Japan*. Stanford: Stanford University Press, 1959.

44 Herbert Spencer, *The Principles of Sociology*. 3 vols. New York: Appleton, 1896-1897.

45 J. W. Spuhler, editor, *The Evolution of Man's Capacity for Culture*. Detroit: Wayne University Press, 1959.

46 Julian H. Steward, "Basin-Plateau Sociopolitical Groups," Bureau of American Ethnology, *Bulletin 120*. Washington: U.S. Government Printing Office, 1938.

47 ———— "Evolution and Process," in A. L. Kroeber, editor, *Anthropology Today*. Chicago: University of Chicago Press, 1952, pp. 313-327.

48 ———— *Irrigation Civilizations: A Comparative Study*. Social Science Monographs, no. 1. Washington: Pan American Union, 1955 (a).

49 ———— *Theory of Culture Change*. Urbana: University of Illinois Press, 1955 (b).

50 ———— "Evolutionary Principles and Social Types," in Tax and Collender, pp. 169-186.

51 Julian H. Steward and Louis C. Faron, *Native Peoples of South America*. New York: McGraw-Hill, 1959.

52 Sol Tax and Charles Callender, editors, *Evolution after Darwin*. The University of Chicago Centennial. 3 vols. Chicago: University of Chicago Press, 1960.

53 Richard Thurnwald, *Die menschliche Gesellschaft in ihren ethno-soziologischen Grundlagen*. Berlin-Leipzig: W. de Gruyter & Co., 1931-1935.

54 Edward B. Tylor, *Anthropology*. The International Scientific Series, vol. 62. New York: Appleton, 1896.

55 ———— *Primitive Society*. 2 vols. London: John Murray, 1920.

56 John von Neumann, *The Computer and the Brain*. New Haven: Yale University Press, 1958.

57 Anthony F. C. Wallace, editor, *Selected Papers of the Fifth International Congress of Anthropological and Ethnological Sciences*. Philadelphia: University of Pennsylvania Press, 1960.

58 Leslie B. White, *The Science of Culture*. New York: Farrar, Straus, 1949.

59 ———— *The Evolution of Culture*. New York: McGraw-Hill, 1959 (a).

60 ———— "The Concept of Evolution in Cultural Anthropology," in Meggers, pp. 106-126 (b).

WALTER A. ROSENBLITH

On Some Social Consequences of Scientific and Technological Change

IN MODERN TIMES, advances in technology have been held to provide the most certain basis for the improvement of the human condition and to contribute the most solid substrate of progress. But, as technology came to affect more of human existence, it became apparent that technology would in turn pose new and serious problems for man. In numerous situations the combination of technological developments and considerations of economics or national defense seemed to conflict with the welfare of individuals or even of mankind as a whole.

Such conflicts of technology with human welfare or human institutions have demonstrated that technology in its traditional relation to the physical sciences could not provide an adequate basis for decision-making in matters that concern the use of technology in organized societies. An all-pervasive technology often requires much basic knowledge in the life and social sciences, in addition to what has been accumulated so successfully in the physical sciences. These needs are reflected in current views of what constitutes a sound education for a technological career. Technology is in a state of transition: the strands that represent applied mathematics, physics, and chemistry are becoming intertwined with those that involve bio-medical and behavioral techniques. These latter, however, often relate quite directly to the cultural practices and ideological beliefs of a society.

Societies—whatever their political form—have come to consider technology as their most powerful and universal tool in the planning of their own future. Given the large role that technological resources and potentialities play in contemporary science, the future of science itself depends increasingly upon technology. Since there is hardly an aspect of basic scientific knowledge that cannot be exploited for social purposes, science, technology, and society have become tightly coupled to one another. This tight coupling is one of the dominating characteristics of the countries that are economically and militarily most powerful. It forces scientists, educators, economists, and even moral philosophers and politicians to assess the ways in which science, technology, and society interact at different stages of economic and political development. Hence the many attempts to foretell how rapid technological change is likely to affect the future of different societies and of the various sciences. In a sense, the acquisition of scientific knowledge, which used to be the product of the more or less planned activity of an individual (operating within the *Zeitgeist*) now depends increasingly upon planning at a societal level. The construction of a giant nuclear accelerator, the size of a nation's budget for health research, or the availability of trained people in given fields of science and technology are examples of the varied aspects of such planning.

We are thus in a period in which there is less laissez-faire in both society and science. It is also a period in our history in which a group of scientists and engineers have been projected into a role for which they have had little special training or much experience. It has been our good fortune that almost all these men (most of them physicists) have to a surprising degree exhibited certain qualities of scientific statesmanship. But no amount of intelligence, no part-time service on the part of well-intentioned and highly sophisticated scientists or engineers, no brilliant improvisations can be expected to provide definitive solutions to persisting problems whose basic variables have yet to be identified.

There is little to be gained from reviewing the extensive literature on the over-all social consequences of scientific and technological progress. The social science of these relations has remained badly underdeveloped: we know as yet little of the basic mechanisms that are at work. Unless one can somehow acquire a new set of relevant data or discover significant relations between technological and social variables, little that is truly novel can be said about so general a topic. And yet the winds of scientific and technological change blow ever harder and hotter. The pressures exerted by technological change, be they military in nature or not, will have a determining influence on the future course of history. These forces, which derive from one of man's most exquisite intellectual achievements, will clearly influence the future progress or even the survival of societies. We thus find ourselves in an unprecedented situation in which we lack the time for dispassionate, uninvolved, retroactive studies.

Given this urgency, we shall proceed in a somewhat clinical manner by presenting several slightly synthetic and somewhat disparate case histories, in order to illuminate certain, perhaps critical facets of these complex interrelations. Whether we can by this approach transcend the position of merely viewing with alarm (or amazement) remains to be established.

Changes in Engineering Education

Our first example, or case history, illustrates some of the changes in engineering education that have taken place during the last quarter-century in this country. During that period the gap between physical science and technology has clearly shrunk. Research in applied science and technological development (R & D) exists now at a scale hardly imaginable in the years before World War II. A great industry has grown up around the systematic exploitation of scientific knowledge. This industry absorbs annually an appreciable fraction of all graduating engineers. If these engineers are to remain useful in a technical capacity for years to come, their training needs a broad scientific base. Since the time span of important change is now considerably shorter than the time span of their working life, they must have an attitude toward technical problems that is flexible enough to enable them to attain the new learning that will be required from them in their later professional life. During these last twenty-five years engineers have also come to realize that the engineering sciences are not enough. What used to be known as "engineering judgement" now must be supported by thorough training in certain aspects of the life and social sciences.

Though university catalogues are perhaps not ideal sources of data, they can legitimately be assumed to reflect significant trends in the education of engineers. It is for this reason that we have

Social Consequences of Change

selected certain corresponding passages from two catalogues of the Massachusetts Institute of Technology for 1938-1939 and 1960-1961 (M.I.T.'s centennial year). The passages quoted are parts of statements introducing the curricula of civil and electrical engineering. In the earlier catalogue, these introductions read as follows:

> The Course in *Civil Engineering* provides a training in the fundamental subjects of science and engineering with particular application to the broad field of design, construction, operation, and maintenance of structures and works necessary for modern industry and civilization. These include bridges, buildings, dams, and other fixed structures; hydraulic and sanitary works for water supply and sewerage, water power, river and harbor improvement; and facilities for transportation by railway, highway, water and air. Although the breadth of Civil Engineering opens up wide fields of service to the members of the profession, its various branches rest upon a relatively compact body of principles.
>
> During the first three years, the Course includes the basic subjects of mathematics and science, and surveying, together with other introductory work in the field of civil engineering. In the fourth year, considerable time is devoted to structural theory and design, foundation problems, and construction methods, subjects which are essential in all branches of the Civil Engineering profession. In addition, the student has a choice of elective subjects which provide an opportunity for specialization in a particular field.
>
> ------------------------
>
> Great importance is attached in *Electrical Engineering* to the study of mathematics, chemistry, physics and applied mechanics in the earlier years, and of the theory of electricity and magnetism beginning in the second year and continuing throughout the remainder of the course. The work in Principles of Electrical Engineering is conducted by means of recitations and supervised problem work. Along with these are associated the essential principles of heat engineering, hydraulic power engineering, and economics. The electrical engineering instruction of the third and fourth years takes on a distinctly scientific character besides offering a variety of alternative subjects involving the applications of electricity to the various problems in railroad work, power-station design, electric machine design, industrial applications of electric power, illumination, communications, electric insulation, etc.

Twenty years later we read:

> *Civil and Sanitary Engineering*
>
> Among trends that point the way to the future are the rapid tempo of scientific and technological advance, the explosion of population, and the

rate at which we are plundering our natural resources. Closely associated is the rate at which man seems to be poisoning—or in some instances strangling—the environment in which he lives. Radioactive wastes, water shortages, smog, obsolescent structures are problems that must be faced by civil engineers. Such problems have two common denominators: they all deal with human needs on a large scale; and they all deal with man's environment. Civil engineering at M.I.T. centers on the fulfillment of human needs through the adaptation and control of the land-water-air environment.

Modern technology, which is partly responsible for some of our great environmental problems, also provides powerful tools to yield solutions to these problems. Civil engineering at M.I.T., in bringing these tools to bear, is the instigator of significant change in the engineering of man's environment. As such, it is of particular interest to young men who are not only challenged by the intellectual rigor of science and mathematics but who are also motivated to exploit the frontiers of science and technology on a large-scale basis for the direct benefit of people.

Starting from a relatively modest beginning about one hundred years ago, the field of *electrical engineering* has become one of great variety and scope. Today its products and services influence the daily living of most of the world's population, and the tempo of growth and change continues. Because new concepts and developments are steadily appearing in the field, professional study and training must give more than a competence to handle the problems of the present. Students should not merely seek to acquire specific skills and methods; they must search for deep understanding of their work and direct it toward the future.

Electrical engineering at M.I.T. supports, in the main, two impelling desires of man. The first is his desire to establish communication—between man and man, between man and machine, and between machine and machine. Radar, radio and wire telephony, measurement, electronics, computation, and control—which are broadly referred to as processing and transmitting information—are the fields in which this communication takes place. The second is man's desire to substitute machines for human muscle—the giant motors, machines and power systems that control the wheels of industry and also give us heat and light. Here the issue is to exploit energy for energy's sake, and we refer to these activities as processing, transmitting, and controlling energy.

Here, then, is a special view of a much broader challenge: the education of today's citizen for survival in a world in continuous, rapid change. As the gap between science and technological application narrows, as scientific activities become more and more dependent

Social Consequences of Change

upon vast technological supporting structures, it becomes harder for the nonspecialist to tell where science leaves off and where technology begins. Lacking a knowledge of the underlying scientific principles, the average citizen is disoriented not only with respect to science but also in his attitudes vis-à-vis the ever-changing gadgets and tools that technology showers upon him. It is this situation that reflects itself in Eric Ashby's article "Dons and Crooners"[1] and in C. P. Snow's essay on the two cultures.[2]

Science, Technology, and Foreign Policy

Our second example is focused on certain aspects of the coupling of science, technology, and foreign policy. The evidence here presented comes from a document of the United States Senate Committee on Foreign Relations, *Possible Nonmilitary Scientific Developments and Their Potential Impact on Foreign Policy Problems of the United States*.[3] Excerpts from the authors' summary follow:

Scientific developments in the next decade will give rise to or intensify many problems that must engage the attention of foreign policy planners. Scientific developments will also help solve foreign policy problems. But the outlook is that the progress of science and technology will do more to create or intensify than to ameliorate such problems, unless deliberate policy measures are taken.

Significant developments affecting international communications will result in the next decade from continued advances in physical techniques and facilities and also from advances in the sciences of human behavior relating to mechanisms of individual or group motivation and to meaningful interchange between members of diverse cultures. Developments such as low-cost mass communication devices, translating machines, and teaching machines will provide new opportunities to diffuse knowledge and ideologies. Brain chemistry and the study of brain mechanisms may open up powerfully beneficial and yet potentially dangerous means to control minds.

These developments will be available to both totalitarian and democratic societies. A problem of increasing importance will be to decide how the United States can use the new possibilities offered by the sciences related to communication in ways consistent with democracy and concern for human freedom.

The national interest requires a more conscious direction of scientific activity in ways likely to assist in the achievement of America's international goals. The security and well-being of the United States call for a reap-

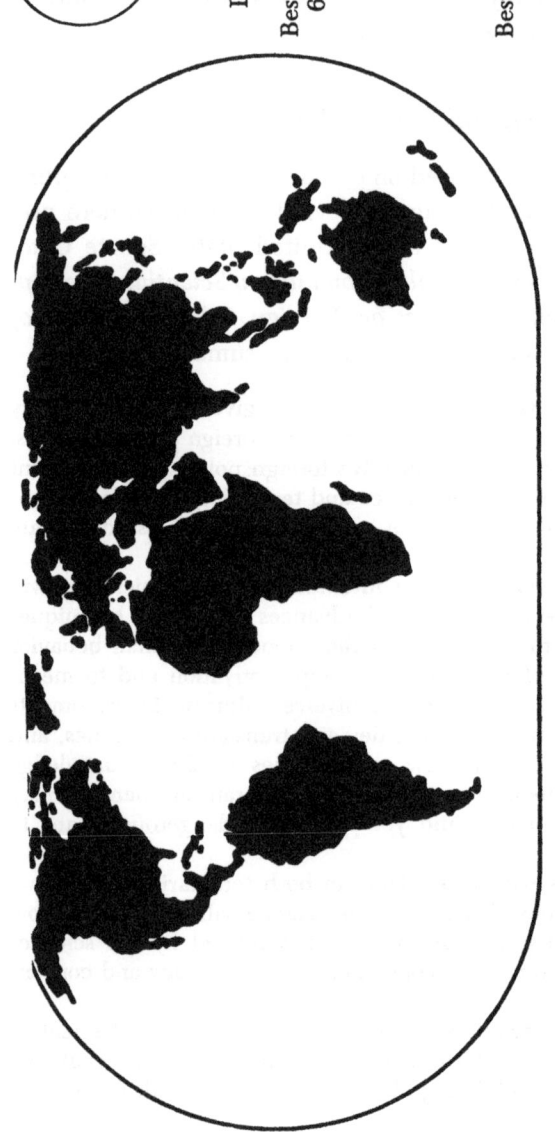

Coal-Steam Railroad and Steamship
Late Nineteenth, Early Twentieth Centuries
Best Regular Speed on Land 65 mph, on Sea 36 mph

Propeller Aircraft to 1950's
Best Regular Passenger Speed 300 mph

Jet Aircraft from 1950's
Regular Passenger Speeds 500 mph and up

Stagecoach and Sailing Ship to 1830's and '40's
Best Regular Speed on Land and Sea 10 mph

SIZE OF THE WORLD, IN TRAVEL TIME
Supposing the Best Travel Technology of Each Epoch Were Applied over the Whole Surface of the Earth

Source: reprinted by permission of the International Industrial Development Center, Stanford Research Institute, from *Possible Nonmilitary Scientific Developments and Their Potential Impact on Foreign Policy Problems of the United States*, Committee on Foreign Relations, the United States Senate, Study No. 2, p. 8.

Social Consequences of Change

praisal of present allocations of scientific and technological effort with a view to directing more effort toward nonmilitary foreign policy challenges.

The present allocation of scientific research and development efforts in the United States and in the world at large is overwhelmingly toward military needs and toward the needs of industry in technologically advanced countries. The emphasis is also overwhelmingly on the physical sciences and their applications, with the biological sciences and their applications second and the psychological and social sciences and their applications very far behind in allocations of effort and support. In view of the fact that foreign policy problems are mainly problems in human relation, this gross imbalance in research effort should be given thoughtful attention.

If the United States is to use science in a more positive manner to support its foreign policy, policy planners will need access to and the aid of the scientific community, including that part of the scientific community concerned with psychological and social disciplines. Planners will need improved methods of analyzing future scientific, political and economic possibilities, in order to create within the necessary lead time the most appropriate policy alternatives to meet the rising tempo of new conditions.

The document expresses agreement with the widely accepted view which holds that science has created problems in foreign policy through the intermediary of technological advances. The chart shows in graphic form the shrinkage of our globe during the last hundred and fifty years. Were we to graph in comparable fashion the time required for the delivery of messages between states, the speed and the range of weapons, we might be able to assess contemporaneous changes in the techniques of diplomacy (from the Congress of Vienna to the United Nations and itinerant summitry) against a set of technologically sensitive variables.

The United States, of course, is not alone among the nations in trying to integrate science and technology into the planning of its foreign policy. This generation of scientists, engineers, and statesmen must face such questions as the following. Can this globe go on divided into a few scientifically and technologically affluent nations, while most nations continue to be underdeveloped in this respect? Can science and scientists in the countries with advanced technology be left entirely free to fulfill their own needs and to follow their own paths?

Last summer, the Rehovoth (near Jerusalem) conference on science and the new states dealt with these very problems. There gathered at the Weizmann Institute statesmen from many countries in Africa and Asia, as well as eminent scientists from the Western world. Together they inquired into ways of making modern science and technology responsive to the needs and aspirations of countries that have only recently gained their independence. The deliberations of the Rehovoth conference deserve to be studied in detail by those concerned with the effects of scientific and technological progress upon human welfare.

There was much concern with the kind of scientific activity that would be most appropriate for these emerging states. W. Arthur Lewis (the principal of University College of the West Indies) expressed a soberly realistic view: [4]

> Fortunately for the new states, they do not have to be in the forefront of developing new science. What they need is rather the application to their problems of what is already known. This is science at the humdrum level, not to be compared with the glamorous and exciting games with the infinite. Most of all, those countries need to survey their physical resources to study the rocks and minerals and soils, rainfall, river-flows, and underground water, fisheries and forests, and to turn to the biological sciences for help in agriculture, in breeding new plants and animals, devising controls of pests and diseases. Since three-quarters of the people, who are also poorly nourished, are engaged in agriculture, this is perhaps the biggest contribution which science can make to their well-being.

The Capabilities of the Computer-Aided Human Brain

Our third case history touches upon a technological development that is likely to become one of humanity's most important scientific and social tools. The invention and the use of tools has enabled man to enhance the power of his muscle, to shape and control his environment. By means of scientific instruments, man has extended and refined his sensorium. His cultural heritage has furnished him with other tools, some of which took the form of theories and experimental apparatus. Thus the intellectual potentialities of today's students

Social Consequences of Change

consist of the accumulation of the great scientific insights of the past.

We are now entering into an era in which the revolutionary technology of computers is about to provide us with a new intellectual tool that will result in an amplification of the capacities of the human brain. The computer will not only constitute a labor-saving device and bring about a new division of intellectual labor, but the combination of man plus computer[5] bids fair to be the equivalent of a jump in the evolution of man's nervous system.

The number of scientists at work today and the undeniable fact that so many among them feel impelled to investigate complex multivariate systems have led to a situation in which many sciences are becoming "data-rich." (Whether the data being collected are the "right" kind is not at issue here.) Cosmic-ray data, plates from nuclear accelerators, X-ray data from organic molecules, electrophysiological data from the nervous system, epidemiological data, data from longitudinal and cross-cultural studies—all seem to cry out for appropriate ways of being handled. The processing of these masses of data, their mere absorption by human nervous systems, call for the use, and more particularly for the refinement, of techniques that are more commonly associated with data from business transactions, census findings, or vast engineering projects. Thus the need exists for a set of flexible computing machines that can be readily programed to carry out specifiable operations that will reduce the amount of information to be handled by scientists. Sometimes such data processing will involve long calculations designed to test highly intricate theoretical concepts; on other occasions the processing may have as its aim the pulling of "signals"* out of a background of "noise."* These same computing machines should also enable a scientist to test rapidly certain hunches he has concerning his data. They should finally help him to deepen his understanding of certain phenomena by permitting him to simulate these events on the basis of a plausible model.

Such uses of computers involve little that is not available now, though as yet few scientists have the requisite skills to realize to the fullest the performance of the new symbiotic unit, man plus computer. There is little immediate need, therefore, to enter either into

---

* Both terms are used in their literal as well as their metaphorical sense.

the realm of science fiction or into arguments as to whether machines *really* think or learn, whether they are likely to build machines in their own image, or will take over our social and political institutions.

There can be little doubt that computers will profoundly affect the ways in which we learn, think, build other computers, and plan the operations of our society. This type of influence is in principle nothing new: the style of science and the strategy of scientists have always been affected by the technological means at their disposal. There has been a tendency to underestimate the effectiveness of the technological feedback upon science, because of the widely held view according to which technology consists primarily of the application of known scientific principles. Just as the microscope, the telescope, the vacuum pump, and the galvanometer helped open up new areas of research, the technology of computers and automata—for whose behavior we have as yet no general theory—makes it possible to deal with scientific problems that we previously had to consider as being outside our reach.

The technology of "electronic brains" presents no particular mystery as long as we focus our attention on the basic properties of these machines: their ability to carry out elementary logical operations reliably in millionths of a second; their ability to store large amounts of data in a standardized form in which access to these data can again be achieved in millionths of a second; input and output devices capable of injecting and extracting the information in appropriate ways; finally, the entire facility must be flexible enough in its programing so that it can carry out a broad spectrum of analytical procedures, not just a single one.

At present, a powerful computer installation—no matter how small the componentry—still demands a great deal of space, a specialized physical environment, and a specialized maintenance personnel. In a few years, however, such giant installations may have many fewer specialized requirements; they may also be capable of dealing quasi-simultaneously, i. e., on the basis of "time-sharing," with a variety of problems without confusing them. Probably also, the time is not too far distant when new materials (that molecular engineering permits us to synthesize according to specifications [6]) will make it possible to build portable computers that a scientist could carry in his rucksack into the field. Such a portable computer might come equipped with a communication link. It would then be possible to enter into instant contact with a powerful installation capable of checking references, other data, or even hypotheses, as soon as they are formulated.

As man's life span lengthens, as he is forced to adapt to continuous change, he will need new techniques of learning. Specialized devices such as teaching machines, together with a vastly increased knowledge of the mechanisms of learning, may reconstruct our whole educational process. There is no reason to assume that computers of the portable kind will not come to play the role of prosthetic devices in relation to the central nervous system much like the role hearing aids and glasses play in relation to the special senses.

Such developments would put us within reach of a technology consonant with Fourier's view of mathematics, as expressed in the preface to his treatise, *The Analytical Theory of Heat*:

> Mathematical analysis is as extensive as nature itself; it defines all perceptible relations, measures time, spaces, forces, temperatures. . . . Its chief attribute is clearness; it has no marks to express confused notions. It brings together the most diverse phenomena, and discovers the hidden analogies which unite them. . . . It seems to be a faculty of the human mind destined to supplement the shortness of life and the imperfection of the senses.

Some Data on the Growth of Science

The final case history aims at a sketchy summary of the growth of the scientific enterprise. The figures given here come from a variety of not entirely compatible sources and can certainly lay no claim to great accuracy. As happens so often with this type of data, absolute amounts are much less reliably estimated than are relative percentages or trends.

At present, most measures of scientific activity indicate that the "doubling time" lies somewhere between 10 and 15 years, while other human activities double approximately every 40 years.* Perhaps the most tangible evidence for this estimate comes from the increase in the number of scientific societies and journals; these

---

* The population of our globe is currently expected to double in less than forty years.

numbered about 100 at the beginning of the nineteenth century, reached 1,000 in 1850, more than 10,000 in 1900, and exceeded 100,000 in 1950. If the doubling interval for science lies indeed in the range between 10 and 15 years, then—as has been pointed out— 80 to 90 percent of all the scientists who have lived since the beginning of history are alive today.

A recent UNESCO report[7] estimates that the number of research workers now alive is approximately 2,000,000. The following figures outline the present composition of the scientific and technological community in the United States:

"High status"* scientists:	approximately	100,000
"Acceptable"* scientists:	approximately	200,000
Engineers:	approximately	650,000
Physicians:	approximately	250,000
Science teachers:	approximately	150,000
Technical assistants:	approximately	1,000,000

For approximately this same period, the figures for scientific workers and physicians in the Soviet Union are of the same order of magnitude, though in much more rapid state of growth.

During the late 1950's, 95,000 engineers graduated annually in the Soviet Union. In the United States, 35,000 students each year obtained bachelor's degrees in engineering, as did approximately the same number in the sciences. The statistics for advanced degrees in the United States are as follows: 5,000 students obtain master's degrees in engineering, while 6,000 students obtain master's degrees in the sciences; finally, there are awarded annually about 3,500 doctor's degrees in the several fields of science and engineering.

A few data should be cited on research costs and research support. The same UNESCO report estimates that if the cost of development is 100, the cost of pure research is 1, the cost of guided fundamental research 3, and of applied research 6. In the United States, the research and development cost has gone up from 300 million dollars in 1938 to roughly 12½ billion dollars in 1959-1960. Of this total, approximately 8 percent have been estimated to support so-called basic research. Of every dollar from the federal-government budget for research in 1957-1958, 31 cents went to the life (the bio-

---

* These and other classifying and defining terms are obviously rather arbitrary.

logical, medical and agricultural) sciences, 64 cents to the physical sciences (including research in engineering), and 5 cents to the social sciences.

The above figures, inaccurate and sketchy as they may be, suffice to convey the impression that science is a rapidly, often exponentially, growing enterprise. In all industrial societies, science is already "big" or likely to become so, in terms of national activities.* In undergoing a significant degree of institutionalization, science and technology (especially science) have changed in character. At the frontier of research only some intellectual activities have been much affected, while scientific-technological massive assaults for "breakthroughs," such as the Manhattan project, are still infrequent. Yet throughout the entire realm of science new organizational arrangements, new practices, new temperaments and new motivations are proliferating. It is, of course, perfectly possible for scientists to decide by fiat that these new-fangled trappings are not science, that as a matter of fact they have nothing to do with science, and that the term "science" applies only to those activities that have undergone relatively few changes since the times of Newton and Maxwell. It is questionable whether at this stage of the history of science such an attitude is realistic in terms of the social role that has devolved upon science and its practitioners.

Scientists may bemoan the fact that most of the people who make political or other important social decisions in this age are in their training and outlook "prescientific," even though they may have learned to consider science as an important military asset. By and large, however, scientists have refused to enter in sufficient numbers and for sufficiently prolonged periods into the social and political process. Instead, they have too often looked askance upon the efforts of those scientists who have done so.

The Implications

Among the representative bodies of American scientists who deal with science and public issues, none takes a broader or more serious

---

* The following corollary is suggested without proof: the social importance of science is much greater than its proportionate share of the gross national product.

view of the social responsibilities of scientists than does the American Association for the Advancement of Science. Its Committee on Science in the Promotion of Human Welfare has recently[8] restated some of the basic issues, with emphasis upon certain new and disturbing features:

What, then, is the scientist's responsibility to his own nation's scientific effort? Clearly, we need to understand that what science contributes to the national purpose is measured by what it adds to the sum of human knowledge; science serves the nation by serving humanity.

A further examination of the effects of the present social uses of science on life inside the house of science itself leads to even more disturbing conclusions. There is some evidence that the integrity of science is beginning to erode under the abrasive pressure of its close partnership with economic, social and political affairs.

The committee suggests a course of action for the scientific community, which, "as the producer and custodian of scientific knowledge ... has the obligation to impart such knowledge to the public." Imparting this knowledge involves scientists with the entire educational process, and it almost certainly calls for the invention of new forms of adult education. It seems unrealistic, however, to expect that even the most effective education could extricate the whole of science from the compromising partnerships referred to above. Given the growth of science and its coupling with society via technology, it has been inevitable that science should become increasingly entangled with human affairs. In response to this process, a certain amount of differentiation in the scientific community has taken place. This differentiation has led to a multiplication of science-related, pioneering activities.

Thus, we observe that there is *the* frontier, at which new knowledge is being sought, there is a frontier in education, and a frontier at which the use of science is being planned at local, national, and international levels. Ideally, scientists may wish to participate in action at all these frontiers, but such uniformity is in no way obligatory. One does not become a scientist without having participated in the search; and a relatively small number of scientists must be free to pursue the unlimited search without being made to account for it or being made to feel that they are shirking their obligation to society.

Social Consequences of Change

Now that technology has come to play such a dominant role in our lives and the science-technology gap is so narrow, scientists are aware of the increasingly tight coupling between the search and the end product. The more socially conscious among them feel that the ability to predict and control implies a responsibility to advise society on the effects of the products of science. In former years, when these products affected human welfare less dramatically, this responsible concern was largely left to the engineers. Now society needs a large number of public servants, experts in all the sciences and technologies, who will—in part with the aid of the technology of weapons systems—"war game," for the benefit of mankind, the consequences of new knowledge and of technological capabilities. Only such a commitment by all social groups, including the scientific-technological community, can bring about a more rational use of scientific and technological change on behalf of human welfare and for the sake of peace.

References

1 Eric Ashby, "Dons and Crooners," *Science*, 1960, *131*: 1165-1170.
2 C. P. Snow, *The Two Cultures and the Scientific Revolution* (The Rede Lecture). Cambridge-New York: Cambridge University Press, 1959.
3 This study was prepared by the Stanford Research Institute; E. Staley and G. Benveniste were the chief authors. It was later followed by another, *Developments in Military Technology and Their Impact on United States Strategy and Foreign Policy*.
4 Quoted from R. Calder, "Science and the New States," *The New Scientist*, August 1960, *8*: 526-527.
5 J. C. R. Licklider, "Man—Computer Symbiosis," *IRE Transactions on Human Factors in Electronics*, March 1960, *1*: 4-11.
6 A. von Hippel, *Principles of Modern Materials Research*. Technical Report No. 136, Laboratory for Insulation Research, Massachusetts Institute of Technology, March 1959.
7 P. Auger, *Survey on the Main Trends of Inquiry in the Field of the Natural Sciences, the Dissemination of Scientific Knowledge and the Application of Such Knowledge for Peaceful Ends*. The United Nations Economic and Social Council, E / 3362, 13 May 1960; Derek J. Price, "The Exponential Curve of Science," *Discovery*, 1956, *17*: 240-243; and Paul Weiss, "Knowledge, a Growth Process," *Science*, 1960, *131*: 1716-1719.
8 "Science and Human Welfare," *Science*, 1960, *132*: 68-73.

Comments on Cultural Evolution

THE COMMENTATORS on cultural evolution are: George G. Simpson, Professor of Vertebrate Paleontology, Museum of Comparative Zoology, Harvard University; Ralph W. Gerard, Professor and Director of Laboratories, Mental Health Research Institute, University of Michigan; Ward H. Goodenough, Professor of Anthropology, the University of Pennsylvania; and Alex Inkeles, Professor of Sociology, Harvard University.

G. G. Simpson: Comparisons of organic and cultural evolution have often been made. Many, indeed most, of these discussions have betrayed ignorance of the comparison, on one side or the other, and naïveté in evaluating the relations between the two sides. The contribution to this subject by Steward and Shimkin is particularly welcome, because it is neither ignorant nor naïve. With the single exception of rejecting any possible specific genetic effects on behavior, it is a well-informed and judicious evaluation of the situation. Nevertheless, there are many points on which more should be said, and among them these three may be stressed:

1. organic evolution is not orthogenetic and there is no *scala naturae;*

2. cultural evolution occurs within a single biological species;

3. there is a real biological and not merely an analogical relation between organic and cultural evolution.

There is nothing new about these observations, but they were either overlooked or misunderstood in much previous discussion of cultural evolution.

As Steward and Shimkin have shown, nineteenth-century theoretical schemes of cultural evolution sought or assumed a single sequence of progressive stages through which any and all cultures were supposed to evolve. This implied a concept of evolution in general, including organic evolution, as tending to change in a rectilinear fashion ("orthogenetically," in later biological jargon) and as tending to follow a single progressive sequence (a reinterpretation of the old *scala naturae*). That conception of organic evolution was the central point of Lamarck's theory, to which the inheritance of acquired characters was only incidental, and indeed its origins were in ancient and pre-evolutionary philosophy. It was implicitly contradicted by Darwin, and, in spite of occasional arguments over orthogenesis, it has been almost completely controverted and abandoned today. As a broad picture of evolutionary tendencies, it is replaced by adaptive radiations diverging into different, available ecological situations and by the periodic replacement of one group by another within any one such situation. It certainly does not follow that analogous processes necessarily occur in cultural evolution, but it is clear that the premises of the theoretical formulations of the nineteenth-century cultural evolutionists are not the only ones possible, that they derive no support from analogy with organic evolution, and that, indeed, they are not valid for organic evolution.

Most evolutionists today are convinced that the same factors are determinative for all phases of evolution and that so-called macro-evolution is only the additive result of so-called micro-evolutionary events. Nevertheless, the interplay of the various factors and the significance of each are relative to circumstances and to the levels at which the phenomena are studied. For example, genetic recombination has certainly underlain the origin of higher categories among most organisms, but the levels at which it actually occurs are only those of single populations and groups of closely related populations. Many of the principles, true or false, for which applications or analogies have been sought by cultural evolutionists were based mainly on supraspecific aspects of organic evolution. Some of them have little or no pertinence for infraspecific populations. This is true, for instance, of such factors or phenomena as adaptive radiation, interspecific isolation, irreversibility, or (a false factor in its usual formulation) orthogenesis. *Homo sapiens* is both genetically and behaviorally a single species. Apart from any other questions of analogy and interpretation, concepts primarily bearing on supraspecific evolution are doubtfully or not at all applicable to evolution within this one species. On the other hand, such factors as recombination, local adaptation and subspeciation, gene migration, intraspecific selection, and others are definitely pertinent to organic evolution within our single species. Whatever relation there may be between organic and cultural evolution should be sought in this field.

From the biological point of view, culture is simply the behavior—or part of the behavior—of a particular species of primates. In principle it can be studied as animal behavior and interpreted in the same ways as the behavior of any other species, for instance, in its adaptive aspects and consequent interaction with natural selection. This approach, involving a real and functional biological relation, may prove to be far more enlightening than seeking strained and dubious analogies between man's biological and cultural evolutions, as if those were two quite separate phenomena. To be sure, human cultures are unique evolutionary developments in numerous respects. Their most important biological peculiarities may be subsumed in these three ways: (a) the content is different, including elements not present (at least not in the same form) in any other species; (b) the behavior involved is extraordinarily varied and complex; and (c) it is exceptionally labile.

The first point merely means that much human behavior (by no means all of it) is species-specific. A great many other animals also have species-specific behavior. Interesting problems arise in relating such behavior to the history and the adaptive status of the particular species in question, but the mere fact that man does have species-specific behavior would seem to demand no different or nonbiological approach to its study.

The complexity of cultural behavior, (b), certainly adds to the descriptive task but again does not obviously demand any difference in approach or principle from that involved in biological study of simpler behavior in other species.

It is the third point, (c), the lability of cultural behavior, that does raise really serious questions about a strictly biological behavioral approach to the study of culture. It is almost obvious, and is supported by quantities of evidence, that many differences in individual human behavior and in human cultures are mainly or completely phenotypic, almost or wholly independent of the genotype. Some psychologists, as regards individual behavior, and many ethnologists, as regards cultural traits, have therefore concluded that for their purposes human genotypes can be taken as a constant and ignored in interpreting behavioral and cultural differences. In dealing with such differences, then, it has been held that nonbiological principles must be found. Hence, of course, comes the attempt to deal with organic and cultural evolutions by analogical confrontation only, rather than in terms of possible real biological relations.

It is a mistake, however, to assume that what is not genetic is not biological, nor is this merely a semantic quibble. Even the geneticists now understand that organic evolution is not synonymous with or necessarily exactly determined by genetic change. Beyond that, the biological nature of an organism is not confined to its genotype, and genetics is not a synonym of biology. Moreover, there are two major errors in attempting to divorce cultural change from genotypic variation. In the first place, although human genotypes in most respects have wide phenotypic behavioral reaction ranges, they do limit behavioral and cultural possibilities, and their reaction ranges are clearly not identical among individuals or among cultural groups. As has been emphasized by others in these conferences, knowledge of human genotypes is extremely poor. It is nevertheless obvious that in highly differentiated cultures like our own some individuals are more and others less well fitted genotypically for certain roles, such as those of university professors or of professional basketball players. Such differences between cultural groups are less clear, but I venture that a population of Eskimos would have not only phenotypic but also genotypic impediments against adopting the integral culture of a tribe of tropical pygmies.

In the second place, cultural traits that are not themselves genetically determined as such may nevertheless affect genetic evolution. As a rather crude but conclusive example, two groups may genetically be equally capable of exploiting a given environment, but the group whose culture does in fact exploit those resources more effectively is likely to increase relative to the other. Thus, natural selection will

COMMENTS

favor one of the groups, which in other respects may be genetically distinct. It is also evident that purely cultural attitudes may have a strong influence on reproductive rates—and differential reproduction *is* natural selection.

I do not question that organic and cultural evolutions can and in some respects must be studied as separate phenomena. The point is that they do interact, that in considerable measure they are parts of a *single* biological phenomenon, and that their study from this point of view is worth while. That study has, in my opinion, been comparatively neglected and it seems to me more promising than further elaboration of merely (often questionable) analogical comparisons between the two subjects as separate fields of enquiry. It is even possible that the study of culture as biological adaptation might provide the general theoretical formulation which, according to Steward and Shimkin, has hitherto been sought with dubious success.

R. W. Gerard: The physiologist is deeply impressed by the problem of individual cells working together in a collective system, so good a system that we often think only of the whole individual rather than of its component members. A point in connection with this collectivity of units, which may have considerable importance to the general theme of these papers, put sharply, is that the greater the collectivity (and this implies both number and integration of units) the more the immediate environment of the single unit is under control, so that the evolution of the whole or of the units in the whole may be shielded or protected from the external environment—what we ordinarily call simply the environment. This is, of course, the warp and woof of physiology: the constancy of the *milieu intérieur* of Claude Bernard, the homeostatic feed-back mechanisms of Cannon, the fact that most of the cells of our body, and most explicitly the neurons of the brain which have far more elaborate homeostatic protections than any other cells in the body, are maintained in a constant immediate environment. They no longer in any sense have to evolve to meet an outside environment. The collective system is evolving to maintain the inside environment constant. I think this is entirely comparable to what culture is doing to civilized man, in whom the individual unit is now evolving very little. Culture or society has an entity, profoundly seen in the collective mechanisms for maintaining constant the conditions in which man must survive and reproduce and all the rest.

"The fixation of experience" underlies evolution, it underlies development, and it underlies learning. I am not talking Lamarckism when I say this, because no system changes as a result of interaction with the surrounding world during its ongoing existence, unless there are changes left in that system from the interaction, unless it is able to fix some of the transactions that have occurred between it and its surroundings. This fixation of experience, which ultimately means guided chemical processes, is, I think, the underlying dimension in all becoming.

Many of the issues raised regarding individuals and populations, and a society as a collection of organisms or a single epiorganism, come into rather sharp focus in the brain itself; there is a large population, over 10 billion neurons, in the human brain, and they act to give a unity. The early evolution of the behavioral capacities of the brain, probably up to the anthropods and the most primitive vertebrates, depended considerably on improved units—neurons and fibers, receptor and effector cells; but there is little further change in these from primitive vertebrates to man. What is very different is the kind and effectiveness of neuron interactions, the circuitry, if you will, and, still more, the sheer number available. Clearly, neurons collectively constitute a new entity.

More units of the same kind, even interacting in the same way, can do things that fewer units cannot do, things that are actually quantitatively different, so, as a man's mind is more than the sum of neuron minds, so the collective mind of mankind is clearly a very different thing from man's individual mind.

There are such extremely stimulating parallels between the organization of men in societies or institutions, the organization of neurons in the brain, and the organization of components in modern computers that the charge so often made, that analogical thought about such parallel systems has not been fruitful, will simply not hold water. My own thinking as a professional neurophysiologist, some of my research, and certainly the way I write and teach in this area have been profoundly influenced by studies done, for example, on organization theory at the social level. I think of the neuron patterns of the nervous system in terms of a table of organization, and raise the question of centralization versus decentralization. I think of synapses as decision points functioning as do human decision-makers, and of the problems of inflow of information and outflow of decision in the nervous system; and much that was rather messy before has come into focus. Conversely, I know that the knowledge and thinking of neurophysiologists have been profoundly helpful to people in the other fields.

A few words in closing on the question of analogy, which has inevitably come up. It is like Margaret Fuller and the universe—one might just as well accept thinking by analogy because that *is* the way we think, that is the way the brain is made. The only problem is whether we use analogy well or stupidly. Was it Claude Bernard who said, "Seek simplicity but mistrust it"? I would say, "Seek generalizations but mistrust them." This is a healthy way to proceed, and the issue is not where the generalizations come from but whether or not they are any good.

Some generalizations did generate from purely biological consideration as to directions of social evolution, and, so far as I know or can tell from the speakers, these generalizations are not stupid. Whether they will hold up, whether they will be ultimate ones, or simply condensations of particular answers which lead to richer and more fruitful generalizations, I do not know. They include the expectation that there will be progressive differentiation of the components of societies; that they will become progressively more dependent on one another; that consequently there will be closer integration of the whole, with stronger forces holding the parts together; that there will be relatively greater centralization and more power of the whole over the individual relative to the influence of individuals over the whole; that there will be progressive superposition of larger and larger units encompassing the simpler ones, larger and larger onion skins so to speak; and that gradients of dominance, of power control, will be present and important.

I hope the main effort of these discussions will be a constructive one. What can we do? Where do we go? What social inventions can we make or catalyze or see the need for and encourage others

Cultural Evolution

to make, in the light of our picture of society, however we may have arrived at this picture? What can we do in the way of furthering the control of the internal environment of society so that the human units manage to survive better and more happily rather than to break down? Let us have artificial selection by artificial evolution at the biological and the cultural level, rather than let nature take its slow and often destructive course.

W. H. Goodenough: A community takes its shape and its institutions are established as a result of the actions of individual people. To understand the mechanisms by which patterns of community life develop and change, therefore, we must consider the individual as well as his community and its institutions. This is another way of saying what has already been said, namely, that the processes going on at one level of organization affect the patterns which emerge at a higher level of organization. This is as true of cultural as of biological evolution.

We are now talking about cultural evolution. One of our major problems is that culture is a vague and rather ill-defined term in anthropology, meaning different things to different people. In one of its meanings, culture has reference to the pattern of life within a community—the regularly recurring activities and material and social arrangements which give a community the semblance of a homeostatic system. When we speak of levels of social complexity or of the transformation of society from hunting and gathering to agricultural forms of organization as aspects of cultural evolution, we are speaking of culture in this sense.

This is quite a different sense from the one implied by the common definition of culture as something which is learned. Here we are talking about the contents of individual human minds. Culture in this sense may be said to include:

1. The ways in which people have organized their experience of their natural, social, and behavioral environment so as to give it structure as a phenomenal world of forms, i.e., their percepts and concepts.

2. The ways in which people have organized their experience of the phenomenal world so as to give it structure as a system of cause-

and-effect relations, i.e., the propositions and beliefs by which they explain events and design tactics for accomplishing their purposes.

3. The ways in which people have organized their experience of their phenomenal world so as to structure its various arrangements in hierarchies of preferences, i.e., their value or sentiment systems. These provide the principles for selecting and establishing purposes and for keeping purposefully oriented in a changing phenomenal world.

4. The ways in which people have organized their experience of their efforts to accomplish recurring purposes into operational procedures for accomplishing these purposes in the future, i.e., a set of "grammatical" principles of action and a series of recipes for accomplishing particular ends. They include operational procedures for dealing with people as well as for dealing with material things.

Thus defined, culture consists of standards for deciding what is, standards for deciding what can be, standards for deciding how one feels about it, standards for deciding what to do about it, and standards for deciding how to go about doing it. People use their standards as guides for all the decisions, little as well as big, which they must make in the course of everyday life. As the members of a community go about their affairs, constantly making decisions in the light of their standards, the patterns characterizing the community as a whole are brought into being and maintained. We have, therefore, two kinds of culture: culture 1, the recurring patterns which characterize a community as a homeostatic system, and culture 2, people's standards for perceiving, judging, and acting. Culture 1, moreover, is an artifact or product of the human use of culture 2.

As I have defined them, individuals can be said to possess culture 2 but not culture 1, which is the property of a community as a social-ecological system. Strictly speaking, moreover, no two persons can be said to have exactly the same culture 2, because every individual in organizing his experience of phenomena, including the actions and utterances of his fellows, necessarily creates his own. But it is also possible to speak of an aggregation of people (a community's members, collectively) as possessing a culture in the sense of culture 2 as distinct from culture 1.

A part of every man's personal organization of experience is his conception of the percepts, concepts, beliefs, values, and operating principles of his fellows, which he attributes to them in order to make sense of their behavior. In this sense of culture, we recognize individual differences and attribute somewhat different cultures to those people whom we know intimately. On the other hand, we do a lot of generalizing from one person to another, so that we develop in our mental map of our social surroundings a series of generalized cultures which we attribute to different aggregations of our fellows. Thus I have in my own mind, in my own private culture, a conception of several different cultures which I attribute to different sets of others. In order to be understood by my fellows, moreover, I use my conception of their culture as a guide for my behavior in dealing with them, making it my operating culture.

Which of the several cultures available to me I select for any given situation depends on the identifications I make. I am thinking of myself now as a scientist, dealing with fellow scientists in accordance with what I conceive to be the culture of my fellow scientists. If I were to find myself working as a member of a ditch-digging crew, I would try to operate with them in terms of the culture which I attribute to them—if I wished to be identified with them—or I might carefully avoid conducting myself in accordance with their culture so as to emphasize my difference from them. In this way, each of us makes choices among what we understand to be the different cultures available to us in a repertoire of cultures in our mind.

In so far as I try to conduct myself according to the standards I attribute to others, others are likely to attribute to me a culture which is in reality a reflection of the culture I attribute to them. As people who have regular dealings with one another try to conform to the cultures they individually attribute to their mutual fellowship, and as they modify their individual conceptions of these cultures in order to increase their predictive value, adjusting their own behavior accordingly, these conceptions increasingly converge. In this way a high degree of consensus can develop both as regards the cultures they individually attribute to their collectivity and consequently as regards the operating cultures which they individually use to guide their mutual dealings. To the extent that there is such agreement among its members, a social group may be said to have a public culture, a culture which its members share and which belongs to them as a group. Corresponding to the four major aspects of culture which I have outlined, a society's public culture consists of the perceptual and conceptual features embedded in the meanings of the vocabulary of its language and other public symbols, a public body of knowledge and beliefs, a set of public conventions, rules, and recipes regarding behavior and operational procedure, and a public value system implicit in the rules of procedure (as in the distribution of rights, privileges, and duties in the group). The fewer the number of social groups within a community, the fewer the number of public cultures which we might expect to find within it; and the more complex the community or society, then the greater the number of disparate public cultures which are likely to obtain and the fewer the number of situations and contexts for which there is a public culture pertaining to the society as a whole.

Now, it seems evident from the background papers of Steward and Shimkin (summarized on pp. 477-497) that they are concerned with evolution and change in culture 1, in the shapes of societies as complex natural systems. But even if this is our interest, we must look to culture 2 in all of its aspects—to what I have called private, operating, and public cultures—for the human processes affecting the course of change in culture 1. Indeed, the relation of culture 2 to culture 1 is not unlike that of genes and gene pools to the phenotypic characteristics of isolate populations. It is sufficiently different, however, to make a serious pursuit of the analogy dangerous. What, then, are some of the processes at the level of culture 2 which govern the shape of culture 1?

First to consider is the role which culture 2 plays as a resource by which people deal with changing circumstances. As their environment changes posing new problems for them, people start doing different things by way of adjustment, sometimes creating radical alterations of culture 1. But in doing so, they have not necessarily made any serious change in culture 2, but are doing what culture 2 already indicates is the appropriate thing under the circumstances. In Truk, in the Caroline Islands, for example, the public culture contains a set of criteria governing a couple's choice of residence after marriage. There are basically two possibilities, residence with the extended family household based on the wife's lineage or with that based on the husband's lineage, and the couple has a right to exercise either choice. The former choice is preferred, provided certain conditions are met, but if the necessary conditions do not obtain in particular cases, residence with the husband's lineage household becomes the proper thing. As conditions affecting people's circumstances in the community undergo change, we expect that the statistical frequency with which one choice or the other is made may fluctuate considerably through time, producing quite different alignments of people, but without there necessarily being any corresponding change in the principles of choice and decision themselves. In time, of course, a radical shift in the patterns of culture 1 resulting from the much more frequent exercise of a particular choice may feed back on people's ideas as to what is preferable, producing a change in their principles of choice and thus a change in culture 2. Nevertheless, while changes in either culture 1 or culture 2 are likely to affect the other, there is no one-to-one relation between them.

The use of culture 2 as a resource for adjusting to changing circumstances brings us to the mental processes of problem-solving and invention. Much of what goes on here seems to involve the breaking apart of complex combinations of principles and ideas and recombining them in ways which people have not previously used but which are consistent with their existing criteria for combining them, with the principles of logic already established in their culture 2. The breakdown and recombination of component ideas and principles is but one kind of change in culture 2—not so much a change in basic content as a reorganization of the same content. Such reorganization of content takes place in all aspects of culture 2, in people's private, operating, and public cultures. Aside from reorganization there is one other way in which a person's private culture may change, namely, by the development of entirely new organizations of experience which could not be handled within the framework of any existing organization of experience in his private culture, as when he finds it necessary to learn a new language so as to bring order out of confusion in his linguistic surroundings.

Changes in a community's public culture arise in three different ways. In the first place, they result from modifications in the conceptions the community's members have of the cultures which they mutually attribute to one another and which they use as their operating cultures. In the second, the number of contexts for which there is a public culture may increase or decrease, as may also the frequency with which these contexts arise. Finally, change results from an agreement to select a different culture as the model and guide for all from among those in the private cultures of some, at least, of the community's members. The first kind of change is analogous to an alteration of rules for playing football. The second is analogous to changes in the things for which there are no rules at all, increasing or decreasing the freedom of each team to act independently of the expectations of the other. The third is analogous to a decision by the several football teams in the same league to play rugby instead.

Disturbances in this complex process by which a public culture is transmitted and maintained through time can produce more radical changes. Some years ago the small atoll of Sorol in the Caroline Islands suffered a tidal wave which left very few survivors, most of them young people. There were many things in the public culture of the community to which these young people had not yet been exposed and to which they could not now be exposed. The transmission of many ideas and bodies of knowledge was interrupted by this event, with drastic effects upon the public culture of Sorol's people. Not only were the possibilities for Sorol's future public culture altered, so also were the kinds of configurations which were likely to characterize its future culture 1.

Changing conditions and experiences often have a marked effect upon people's private cultures without resulting in an immediate change in public culture. People continue to conduct themselves in accordance with the public culture of their group, because they assume that others expect them to do so. Such is the case when people become aware of new ways of thinking, believing and acting, following contact with members of other communities, and are privately attracted by some of them. At the same time, they continue to conduct themselves in public according to the established public cultures of the groups in their community with which they are identified and to which they believe their fellows still subscribe. Inevitably,

Cultural Evolution

therefore, change in the public culture of a community's members lags behind changes in their private cultures. This lag has the effect of weakening people's commitment to honor the provisions of their public culture, as honoring them becomes less and less fulfilling in the light of their new private wants, values, and beliefs. The weakening commitment produces an increase in delinquency, by the old public culture's standards, and other evidences of "social disorganization." There is increasingly a sense of need for a new, more gratifying, public culture and for a new commitment to abide by its provisions, for what amounts to the third kind of change I mentioned: the deliberate selection of a new public culture.

The Nakanai are a case in point. Their dissatisfaction with themselves following contact with Europeans and their inability to restore their self-esteem through traditional avenues to success in their old public culture have led them to concerted efforts to adopt their conception of European culture as the basis for a new public culture. Such developments obviously contribute to radical changes at the level of culture 1. Indeed, discontent with the existing states of affairs in their community (with its culture 1) may prompt people to look for a new public culture as a necessary means to creating a more desirable culture 1. We see this sort of thing in connection with the nationalistic and economic aspirations of peoples in underdeveloped areas today.

The choices people make as to their operating cultures and new public cultures obviously represent some kind of selective process. It is certainly not natural selection, in the usual sense of that term, but a selection, nevertheless, which has important effects on the course of change in culture 1. If natural selection is an important process or mechanism in biological evolution, psychological selection (if we may call it that) seems to be an equally important mechanism of cultural evolution, directly in the case of changes in operating and public cultures in the realm of culture 2, and indirectly in the realm of culture 1. Identification, moreover, seems to be an important aspect of this selective process.

The processes I have suggested help explain the radiation phenomenon in cultural evolution. We can see cultural radiation reflected in the Pacific Islands. Here there are many local communities whose respective public cultures (culture 2) can be seen as a series of slightly different organizations of and emphases on a number of basic ideas and principles present in all of them. Yet in culture 1, in the over-all patterns which we discern in these communities when we look at them as wholes, there is a much greater degree of difference as we go from one part of the Pacific to another. I think that it is possible to see how these different patterns of culture 1 are largely the products of slightly different emphases in the less divergent underlying cultures of type 2. This is only a hunch that I have about the situation in the Pacific, but it is one which I think could be fruitfully explored by intense empirical study.

Alex Inkeles: Most of what I will say is based on a very thin empirical base. I think, nevertheless, that these problems are worth considering. Concentrating on the modern period, we may say that the sociological evolutionists are substantially agreed that within a period of from twenty-five to seventy-five years most of the people in the world will find themselves in what we call industrial societies. The rapidity with which various people approach this form will vary tremendously, but we think that certainly within the next seventy-five years the transformation will be complete. If we have anything to say about evolution, it consists largely in trying to specify, on the basis of the one model that we have, what most of the world culture will be like as a result of this process of transition from distinctive local cultures to the uniform adoption of the chief elements of the industrial order.

We believe that this process is really distinctive, in the sense that at no other time in history, as near as we can tell, has there been anything like the same uniform spread of certain essential features of a culture pattern and of social structure, especially of features which seem to have very powerful capacity to influence all other aspects of culture. I do not know whether it qualifies as predicting evolution to say that we can foresee the outlines of future societies because we can project into the future onto other peoples the outlines of the industrial society as we know it today. Evolution or not, that is essentially what we are doing. We know some of the internal dynamics or mechanisms of industrial societies. We make a fairly definite assumption that most of the world is going to end up with a system of approximately that kind in the next twenty-five to seventy-five years. Within this framework, we try to suggest what world culture will tend to look like at the end of the next twenty-five to fifty years. We do this mainly by making evaluations of the interrelations between other elements of social structure and the industrial system taken as the main compelling force in the situation. I should like, therefore, briefly to take up the main elements of social structure we generally work with, and to suggest some of our thoughts about what is happening or will happen in each of these realms.

With regard to kinship, the general assumption seems to be that despite the great and rich diversity of the kinship systems which have characterized the world in the past, most of the world will be focused on the isolated nuclear conjugal family as against various forms of extended family systems. In a family system based on the isolated conjugal family, the couple cannot count on a large number of kinfolk for help, as these in turn cannot call on the couple for services. Certain very real consequences in the rest of the system follow from that fact. The kin have few moral or other controls over one another. The implications of that fact for the rest of the system are really very substantial, because this means that people's behavior is not mainly controlled by those to whom they are closely linked in kin systems based upon extensive obligations and rights. In so far as this is the case, somewhere else in the structure we must provide other ways of controlling behavior. The alternatives are either the elaboration of regulations, on the one hand, or, on the other, the increasing intensification of the internalization of social norms.

The next problem is the nature of the community. All our analyses of social life must reckon with a fact as basic as kinship, namely, that of physical relatedness, of the relatedness of people in space. We build up from the kinship unit to larger sets of relations, including the neighborhood or the immediate community, the settlement or village, then the larger spatial or regional connection, and ultimately the national state. To some people, indeed, their idea of community includes a sense of relation to all mankind, depending partly on what is inside themselves, and partly on how they are tied into the network of human relations.

COMMENTS

With regard to the evolution of community, our general expectation is, in the first place, that we will have a progressive and relatively sharp decrease in the proportion of the total population that is widely dispersed over the total area of each country. This is perhaps only another way of saying that rural residence will be replaced by large-scale urban conglomerates. In the process, a great many of the typical patterns of very frequent, intensive, and deeply grooved social relations between physical neighbors will tend to disappear. There is a fairly reliable relation between the level of industrialization and the proportion of the population in cities of a certain size. But there are some countries in which the proportion of the population in industry is relatively low, and yet they boast enormous city areas. These tend to be places in which the rural patterns of life and traditional culture have received a terrible shock. People have been loosened and freed from their primary ties in rural areas and have drifted to cities. On their arrival, however, they do not enter the usual urban employments. They are in the cities simply because social-welfare services in their original rural community have broken down, whereas they know that in the city the society or the state will provide them with social-welfare services, if only to keep them from rioting.

With regard to the stratification realm, sociologists rather disagree about what happens in the industrial society. I put forth my own ideas here as one concrete notion of what will happen, whether right or wrong.

Two major processes apparently occur in the stratification realm as we move from traditional to modern or industrialized society. For one thing, the typical pyramids of stratification tend to become truncated. The tops and the bottoms are cut off, and one ends up with relatively fewer strata. In addition, a larger proportion of the total population in these truncated stratification pyramids is found near the center rather than at the ends. By contrast, in the case of a predominantly peasant community, there is a concentration of people in *one* of the strata, usually the bottom stratum. For example, 80 percent of the Russians were peasants before the Soviet Revolution.

In addition, and I think it most important, there is a tendency toward the homogenization of the conditions of life and the qualities

Cultural Evolution

of people who compose the different strata. Incidentally, our general conclusion is that the form of the stratification pyramid under conditions of modernization comes everywhere to assume the same shape, taking its outlines from the structure of a large-scale, industrial, and bureaucratic organization adapted to urban society. This is in contrast to the nearly unique patterns of stratification often found in the world before the modernization tendency became so widespread.

The economy: with regard to the context of the individual's activities outside kinship, community, and stratification, probably the most important involvement is in economic life, in the social forms which determine how people work, how they produce subsistence for themselves and the rest of the community, in so far as they are involved in an interlocking system of support. The general trend in modern society seems to be toward larger and larger units, characterized by more and more impersonal patterns of interpersonal relations, both between the individual and the materials he works with and between the individual and other individuals in the system of production.

With regard to the polity, the political structure, there is perhaps more uncertainty as to what the main future forms will be than about any other aspect of our problem. There is one very popular view which holds that the whole pattern of industrialization and its consequences with regard to urbanization, impersonalization, and so on leads more or less inevitably to a concentration of planning functions so extensive that something like a totalitarian society is the most likely, indeed, some would argue, the inevitable consequence of the whole process. The weight of opinion at the moment is that there is nothing inherent in the general industrial system which is compelling with regard to either democratic or totalitarian political forms. Industrialism seems to live quite well with both.

With regard to idea systems, two processes are apparently taking place. The diversity of idea systems, characterizing populations at large, is in some respects becoming restricted and in other respects being elaborated. The range or variety of idea systems is becoming restricted in the sense that more and more people in the world tend to share the same pool of idea systems. But there is at the same time an enrichment, in the sense that to more and more people around the world more and more information and skills are available. To take the average individual today, the range of ideas and the pool of information he possesses are much greater than most people around the world possessed at some earlier points in time.

With regard to large-scale systems, the tendency seems to be toward larger and larger units. The national state becomes the basic unit of societal organization, and large states apparently have better long-range prospects than do small nations. We have a lot of evidence that smaller units can exist quite well, and the situation here is not compelling, but there is a great need for innovation, for creativity, in combining high degrees of autonomy for small units, with high degrees of coordination between the small units, so as to make more efficient larger networks.

I am not as pessimistic on the subject of mass culture as are a great many people, including those who reported in the issue of *Dædalus* (Spring, 1960), devoted to mass culture. By and large, we are getting an improvement in the general intellectual quality of people around the world. The information content, the idea content, the skills, the capacity to work, the capacity to deal with and relate to ideas, and to relate to other people over a wide range of situations have increased under the process of industrialization.

Furthermore, I think the general impact on human quality of this whole development has been to increase the probability that the existing but hidden talent in human populations, talent never before developed, will now have a chance to become trained and will exert an influence on the total pool of human knowledge.

One thing industrialism has done, oddly enough, is to increase the general coordination among people and yet to maximize the freedom of any one individual to pursue a relatively independent course of action without being excessively restricted by the requirements of the system as a whole. What we count on is that the system as a whole should produce a certain level of output. We often seem able to get that level of output without the necessity of commanding each unit in the system in very close detail as to what its course of action should be. To some degree, the Soviet Union and the United States represent embodiments of these two alternative modes for organizing a modern industrial society. If we were to take rates of growth as our criterion, we could argue that the Soviet Union is the more adaptive system. If we use the standard suggested for biological evolution, namely, differential reproduction, the Soviet system is perhaps more likely to win out, because it will "reproduce" more, in the sense that more societies are likely to follow the Soviet model than will follow the American model; but I think that so far this is a relatively open question. If you ask, therefore, what is the most important social evolutionary question for the future, it is whether societies of the Western type can meet the challenge of problems such as the rate of growth, without their necessarily copying the structural features adopted by the Soviet world to attain those same ends.

Alexander Lesser (Professor of Anthropology, Hofstra College): Anthropology cannot use Alex Inkeles' crystal ball. Its "consistently comparative point of view" (Kluckhohn) recognizes the major differences in the social, economic, and political institutions that accompanied the industrial revolution in England, Germany, Japan, and Russia, and destroys the illusion that there is only "one model" of industrial society. Studies of ongoing industrialization in underdeveloped areas also prove that political and social institutions of England and the United States are not inevitable concomitants (Colson, Watson, Nash). There is even evidence that historically the individual conjugal family (as against more extended family forms) may have been a forerunner rather than an effect of the industrial revolution in England (Arensberg).

In brief, predictions cannot be based on ethnocentrically limited data of the West and faith in technological determinism. The different and diverse social and political institutions which now exist in underdeveloped areas undergoing or about to undergo industrialization are bound to be factors in the patterns of life which will exist there after industrialization has taken place. Professor Inkeles notes that "industrialism seems to live quite well with both . . . democratic or totalitarian political forms." When industrial technology can coexist with such political extremes, how can we assume and predict that it determines uniformity and identity in family, community, and other social institutions?

B. F. SKINNER

The Design of Cultures

ANYONE WHO UNDERTAKES to improve cultural practices by applying a scientific analysis of human behavior is likely to be told that improvement involves a value judgment beyond the pale of his science and that he is exemplifying objectionable values by proposing to meddle in human affairs and infringe on human freedoms. Scientists themselves often accept this standard contention of Western philosophy, even though it implies that there is a kind of wisdom which is mysteriously denied to them and even though the behavioral scientists among them would be hard pressed to give an empirical account of such wisdom or to discover its sources.

The proposition gains unwarranted strength from the fact that it appears to champion the natural against the artificial. Man is a product of nature, the argument runs, but societies are contrived by men. Man is the measure of all things, and our plans for him—our customs and institutions—will succeed only if they allow for his nature. To this it might be answered that man is more than an immutable product of biological processes; he is a psychological entity, and as such also largely man-made. His cause may be as contrived as society's and possibly as weak. He is, nevertheless, an individual, and his defenders are individuals, too, who may borrow zeal in his defense from their own role in the great conflict between the one and the many. To side with the individual against the state, to take a specific example, is reassuringly to defend one's own, even though it might be answered that mankind has won its battles only because individual men have lost theirs.

These are merely answers in kind, which can no doubt be met with plausible rejoinders. The disputing of values is not only possible,

The Design of Cultures

it is interminable. To escape from it we must get outside the system. We can do this by developing an empirical account of the behavior of both protagonists. All objections to cultural design, like design itself, are forms of human behavior and may be studied as such. It is possible that a plausible account of the design of cultures will allay our traditional anxieties and prepare the way for the effective use of man's intelligence in the construction of his own future.

It is reasonable to hope that a scientific analysis will someday satisfactorily explain how cultural practices arise and are transmitted and how they affect those who engage in them, possibly to further the survival of the practices themselves or at least to contribute to their successors. Such an analysis will embrace the fact that men talk about their cultures and sometimes change them. Changing a culture is itself a cultural practice, and we must know as much as possible about it if we are to question it intelligently. Under what circumstances do men redesign—or, to use a discredited term, reform—their way of life? What is the nature of their behavior in doing so? Is the deliberate manipulation of a culture a threat to the very essence of man or, at the other extreme, an unfathomed source of strength for the culture which encourages it?

We need not go into the details of a scientific account of behavior to see how it bears on this issue. Its contribution must, however, be distinguished from any help to be drawn from historical analogy or the extrapolation of historical trends or cycles, as well as from interpretations based on sociological principles or structures. Such an account must make contact with biology, on the one hand, but serve in an interpretation of social phenomena, on the other. If it is to yield a satisfactory analysis of the design and implementation of social practices, it must be free of a particular defect. Evolutionary theory, especially in its appeal to the notion of survival, suffered for a long time from circularity. It was not satisfying to argue that forms of life which had survived must therefore have had survival value and had survived because of it. A similar weakness is inherent in psychologies based on adjustment or adaptation. It is not satisfying to argue that a man adapts to a new environment because of his intelligence and emotional stability if these are then defined in terms of capacities to adapt. It is true that organisms usually develop in directions which maximize, phylogenetically, the survival of the species and, ontogenetically, the adjustment of the individual; but the mechanisms responsible for both kinds of change need to be explained without recourse to the selective effect of their consequences.

In biology this is now being done. Genetics clarifies and supports evolutionary theory with new kinds of facts, and in doing so eliminates the circularity in the concept of survival. A comparable step in the study of human behavior is to analyze the mechanisms of human action apart from their contribution to personal and cultural adjustment. It is not enough to point out that a given form of behavior is advantageous to the individual or that a cultural practice strengthens the group. We must explain the origin and the perpetuation of both behavior and practice.

A scientific analysis which satisfies these conditions confines itself to individual organisms rather than statistical constructs or interacting groups of organisms, even in the study of social behavior. Its basic datum is the probability of the occurrence of the observable events we call behavior (or of inferred events having the same dimensions). The probability of behavior is accounted for by appeal to the genetic endowment of the organism and its past and present environments, described wholly in the language of physics and biology. The laboratory techniques of such an analysis, and their technological applications, emphasize the prediction and control of behavior via the manipulation of variables. Validation is found primarily in the success with which the subject matter can be controlled.

An example of how such an analysis differs from its predecessors is conveniently at hand. An important group of variables which modify behavior have to do with the consequences of action. "Rewards" and "punishments" are variables of this sort, though rather inadequately identified by those terms. We are interested in the fact (apart from any theory which explains it) that by arranging certain consequences —that is, by making certain kinds of events *contingent upon behavior* —we achieve a high degree of experimental control.* Our present understanding of the so-called "contingencies of reinforcement" is

* To a hungry organism food is a reinforcement. An experimenter "makes food contingent on a response" by connecting the response with the operation of a food magazine. For example, if the response is pressing a lever, the lever may be made to close a switch which operates a magazine electrically. The receipt of food is said to reinforce pressing the lever.

The Design of Cultures

undoubtedly incomplete, but it nevertheless permits us to construct new forms of behavior, to bring behavior under the control of new aspects of the environment, and to maintain it under such control for long periods of time—and all of this often with surprising ease. Extrapolation to less rigorously controlled samples of behavior outside the laboratory has already led to promising technological developments.

But the importance of the principle is embarrassing. Almost any instance of human behavior involves contingencies of reinforcement, and those who have been alerted to their significance by laboratory studies often seem fanatical in pointing them out. Yet behavior *is* important mainly because of its consequences. We may more readily accept this fact if we recall the ubiquity of the concept of purpose. The experimental study of reinforcing contingencies is nothing more than a nonteleological analysis of the *directed effects* of behavior, of relations which have traditionally been described as purpose. By manipulating contingencies of reinforcement in ways which conform to standard practices in the physical sciences, we study and use them without appealing to final causes.

We can put this reinterpretation of purpose to immediate use, for it bears on a confusion between the phylogenetic and the ontogenetic development of behavior which has clouded our thinking about the origin and growth of cultures. Contingencies of reinforcement are similar to what we might call contingencies of survival. Inherited patterns of behavior must have been selected by their contributions to survival in ways which are not unlike those in which the behavior of the individual is selected or shaped by its reinforcing consequences. Both processes exemplify adaption or adjustment, but very different mechanisms must be involved.

The evolution of inherited forms of behavior is as plausible as the evolution of any function of the organism when the environment can be regarded as reasonably stable. The internal environment satisfies this requirement, and a genetic endowment of behavior related to the internal economy—say, peristalsis or sneezing—is usually accepted without question. The external environment is much less stable from generation to generation, but some kinds of responses to it are also plausibly explained by evolutionary selection. The genetic mechanisms are presumably similar to those which account for other functions. But environments change, and any process which permits an organism to modify its behavior is then important. The structures which permit modification must have evolved when organisms were being selected by their survival in novel environments.

Although the mechanisms which permit modification of behavior are inherited, learned behavior does not emerge from, and is not an extension of, the unlearned behavior of the individual. The organism does not simply refine or extend a genetic behavioral endowment to make it more effective or more inclusive. Instead, it develops collateral behavior, which must be distinguished from an inherited response system even when both serve similar functions. It is important to remember this when considering social behavior. In spite of certain intriguing analogies, it is not likely that the social institutions of man are founded on or that they emerged from the instinctive patterns of animal societies. They are the achievements of individuals, modifying their behavior as inherited mechanisms permit. The coordinated activities of the anthill or beehive operate on very different principles from those of a family, a large company, or a great city. The two kinds of social behavior must have developed through different processes, and they are maintained in force for different reasons.

To take a specific example, verbal behavior is not a refinement upon instinctive cries of alarm, distress, and so on, even though the reinforcing contingencies in the one case are analogous to the conditions of survival in the other. Both may be said to serve similar adaptive functions, but the mechanisms involved in acquiring verbal behavior clearly set it apart from instinctive responses. The innate vocal endowment of an organism is indeed particularly refractory to modification, most if not all verbal responses being modifications of a nonspecific behavioral endowment.

In general, the evolution of man has emphasized modifiability rather than the transmission of specific forms of behavior. Inherited verbal or other social responses are fragmentary and trivial. By far the greater part of behavior develops in the individual through processes of conditioning, given a normal biological endowment. Man becomes a social creature only because other men are important parts of his environment. The behavior of a child born into a flourishing society is shaped and maintained by variables, most of which are

The Design of Cultures

arranged by other people. These social variables compose the "culture" in which the child lives, and they shape his behavior in conformity with that culture, usually in such a way that he in turn tends to perpetuate it. The behavioral processes present no special problems. Nevertheless, a satisfactory account calls for some explanation of how a social environment can have arisen from nonsocial precursors. This may seem to raise the hoary question of the origin of society, but we have no need to reconstruct an actual historical event or even a speculative beginning, such as a social compact from which conclusions about the nature of society can be drawn. We have only to show that a social environment could have emerged from nonsocial conditions. As in explaining the origin of life, we cannot discover an actual historical event but must be satisfied with a demonstration that certain structures with their associated functions could have arisen under plausible conditions.

The emergence of a given form of social behavior from nonsocial antecedents is exemplified by imitation. Inherited imitative behavior is hard to demonstrate. The parrot may possibly owe its distinction only to an inherited capacity to be reinforced by the production of imitative sounds. In any case, an inherited repertoire of imitative behavior in man is insignificant, compared with the product of certain powerful contingencies of reinforcement which establish and maintain behaving-as-others-behave. For example, if organism A sees organism B running in obvious alarm, A will probably avoid aversive consequences by running in the same direction. Or, if A sees B picking and eating ripe berries, A will probably be reinforced for approaching the same berry patch. Thousands of instances of this sort compose a general contingency providing for the reinforcement of doing-as-others-do. In this sense, behavior exemplifying imitation is acquired, yet it is practically inevitable whenever two or more organisms live in contact with one another. The essential conditions are not in themselves social.

Most social behavior, however, arises from social antecedents. Transmission is more important than social invention. Unlike the origin of cultural practices, their transmission need not be a matter for speculation, since the process can be observed. Deliberate transmission (that is, transmission achieved because of practices which have been reinforced by their consequences) is not needed. For example, some practices are perpetuated as the members of a group are severally replaced. If *A* has already developed specific controlling behavior with respect to *B*, depending partly upon incidental characteristics of *B*'s behavior, he may impose the same control on a new individual, *C*, who might not himself have generated just the same practices in *A*. A mother who has shaped the vocal responses of her first baby into a primitive verbal repertoire may bring already established contingencies to bear on a second child. A leader who has acquired aversive controlling practices in his interactions with a submissive follower may take by storm a second follower even though, without this preparation, the leader-follower relation might have been reversed in the second case. Overlapping group membership is, of course, only one factor contributing to manners, customs, folkways, and other abiding features of a social environment.

These simple examples are offered not as solutions to important problems but to illustrate an approach to the analysis of social behavior and to the design of a culture. A special kind of social behavior emerges when *A* responds in a definite way *because of the effect on the behavior of B*. We must consider the importance of *B* to *A* as well as of *A* to *B*. For example, when *A* sees *B* looking into a store window, he is likely to be reinforced if he looks too, as in the example of the berry patch. But if his looking is important to *B*, or to a third person who controls *B*, a change may take place in *B*'s behavior. *B* may look into the window in order to induce *A* to do the same. The carnival shill plays on the behavior of prospective customers in this way. *B*'s behavior is no longer controlled by what is seen in the window but (directly or indirectly) by the effect of that behavior on *A*. (The original contingencies for *A* break down: the window may not now be "worth looking into.") Action taken by *A* because of its effect on the behavior of *B* may be called "personal control." An important subdivision is verbal behavior, the properties of which derive from the fact that reinforcements are mediated by other organisms.[1] Another subdivision is cultural design.

In analyzing any social episode from this point of view a complete account must be given of the behaviors of both parties as they contribute to the origin and maintenance of the behavior of each other. For example, in analyzing a verbal episode, we must account for both speaker and listener. This is seldom done in the case of non-

The Design of Cultures

verbal personal control. In noticing how the master controls the slave or the employer the worker, we commonly overlook reciprocal effects and, by considering action in one direction only, are led to regard control as exploitation, or at least the gaining of a onesided advantage; but the control is actually mutual. The slave controls the master as completely as the master the slave, in the sense that the techniques of punishment employed by the master have been selected by the slave's behavior in submitting to them. This does not mean that the notion of exploitation is meaningless or that we may not appropriately ask, *Cui bono*? In doing so, however, we go beyond the account of the social episode itself and consider certain long-term effects which are clearly related to the question of value judgments. A comparable consideration arises in the analysis of any behavior which alters a cultural practice.

We may not be satisfied with an explanation of the behavior of two parties in a social interaction. The slaves in a quarry cutting stone for a pyramid work to escape punishment or death, and the rising pyramid is sufficiently reinforcing to the reigning pharaoh to induce him to devote part of his wealth to maintaining the forces which punish or kill. An employer pays sufficient wages to induce men to work for him, and the products of their labor reimburse him, let us say, with a great deal to spare. These are on-going social systems, but in thus analyzing them we may not have taken everything into account. The system may be altered by outsiders in whom sympathy with, or fear of, the lot of the slave or exploited worker may be generated. More important, perhaps, is the possibility that the system may not actually be in equilibrium. It may breed changes which lead to its destruction. Control through punishment may lead to increasing viciousness, with an eventual loss of the support of those needed to maintain it; and the increasing poverty of the worker and the resulting increase in the economic power of the employer may also lead to countercontrolling action.

A culture which raises the question of collateral or deferred effects is most likely to discover and adopt practices which will survive or, as conditions change, will lead to modifications which in turn will survive. This is an important step in cultural design, but it is not easily taken. Long-term consequences are usually not obvious, and there is little inducement to pay any attention to them.

We may admire a man who submits to aversive stimulation for the sake of later reinforcement or who eschews immediate reinforcement to avoid later punishment, but the contingencies which lead him to be "reasonable" in this sense (our admiration is part of them) are by no means overpowering. It has taken civilized societies a long time to invent the verbal devices—the precepts of morals and ethics—which successfully promote such an outcome. Ultimate advantages seem to be particularly easy to overlook in the control of behavior, where a quick though slight advantage may have undue weight. Thus, although we boast that the birch rod has been abandoned, most school children are still under aversive control—not because punishment is more effective in the long run, but because it yields immediate results. It is easier for the teacher to control the student by threatening punishment than by using positive reinforcement with its deferred, though more powerful, effects.

A culture which has become sensitive to the long-term consequences of its measures is usually supported by a literature or philosophy which includes a set of statements expressing the relations between measures and consequences. To the cultural designer, these statements function as prescriptions for effective action; to the members of the group, they are important variables furthering effective self-management. (To both, and to the neutral observer, they are sometimes said to "justify" a measure, but this may mean nothing more than strengthening the measure by classifying it with certain kinds of events characteristically called "good" or "right.") Thus, a government may induce its citizens to submit to the hardship and tragedy of war by picturing a future in which the world is made safe for democracy or free of Communism, or to a program of austerity by pointing to economic changes which will eventually lead to an abundance of good things for all. In so doing, it strengthens certain behavior on the part of its citizens which is essential to its purposes, and the resulting gain in power reinforces the government's own concern for deferred effects and its efforts to formulate them.

The scientific study of behavior underlines the collateral effects of controlling practices and reveals unstable features of a given interaction which may lead to long-deferred consequences. It may dictate effective remedial or preventive measures. It does not do this, however, by taking the scientist out of the causal stream.

The Design of Cultures

The scientist also is the product of a genetic endowment and an environmental history. He also is controlled by the culture or cultures to which he belongs. Doing-something-about-human-behavior is a kind of social action, and its products and by-products must be understood accordingly.

A reciprocal relationship between the knower and the known, common to all the sciences, is important here. A laboratory for the study of behavior contains many devices for controlling the environment and for recording and analyzing the behavior of organisms. With the help of these devices and their associated techniques, we change the behavior of an organism in various ways, with considerable precision. *But note that the organism changes our behavior in quite as precise a fashion.* Our apparatus was designed by the organism we study, for it was the organism which led us to choose a particular manipulandum, particular categories of stimulation, particular modes of reinforcement, and so on, and to record particular aspects of its behavior. Measures which were successful were for that reason reinforcing and have been retained, while others have been, as we say, extinguished. The verbal behavior with which we analyze our data has been shaped in a similar way: order and consistency emerged to reinforce certain practices which were adopted, while other practices suffered extinction and were abandoned. (All scientific techniques, as well as scientific knowledge itself, are generated in this way. A cyclotron is "designed" by the particles it is to control, and a theory is written by the particles it is to explain, as the behavior of these particles shapes the nonverbal and verbal behavior of the scientist.)

A similarly reciprocal effect is involved in social action, especially in cultural design. Governmental, religious, economic, educational, and therapeutic institutions have been analyzed in many ways—for example, as systems which exalt such entities as sovereignty, virtue, utility, wisdom, and health. There is a considerable advantage in considering these institutions simply as behavioral technologies. Each one uses an identifiable set of techniques for the control of human behavior, distinguished by the variables manipulated. The discovery and invention of such techniques and their later abandonment or continued use—in short, their evolution—are, or should be, a part of the history of technology. The issues they raise, particularly with respect to the behavior of the discoverer or inventor, are characteristic of technology in general.

Both physical and behavioral technologies have shown progress or improvement in the sense that new practices have been discovered or invented and tested and that some of them have survived because their effects were reinforcing. Men have found better ways, not only to dye a cloth or build a bridge, but to govern, teach, and employ. The conditions under which all such practices originate range from sheer accident to the extremely complex behaviors called thinking.[2] The conditions under which they are tested and selected are equally diverse. Certain immediate personal advantages may well have been the only important variables in the behavior of the primitive inventors of both physical and cultural devices. But the elaboration of moral and ethical practices has reduced the importance of personal aggrandizement. The honorific reinforcements with which society encourages action for the common weal, as well as the sanctions it applies to selfish behavior, generate a relatively disinterested creativity. Even in the field of personal control, improvements may be proposed, not for immediate exploitation, but—as by religious leaders, benevolent rulers, political philosophers, and educators—for "the good of all."

Only an analysis of moral and ethical practices will clarify the behavior of the cultural designer at this stage. He has faced a special difficulty in the fact that it is easier to demonstrate the right way to build a bridge than the right way to treat one's fellowmen (the difference reducing to the immediacy and clarity of the results). The cultural inventor, even though relatively disinterested, has found it necessary to appeal for support to secular or divine authorities, supposedly inviolable philosophical premises, and even to military persuasion. Nothing of the sort has been needed for the greater part of physical technology. The wheel was not propagated by the sword or by promises of salvation—it made its own way. Cultural practices have survived or fallen only in part because of their effect on the strength of the group, and those which have survived are usually burdened with unnecessary impedimenta. By association, the current designer is handicapped by the fact that men look behind any cultural invention for irrelevant, ingenuous, or threatening forces.

There is another step in physical technology, however, which must have a parallel in cultural design. The practical application of scientific knowledge shows a new kind of disinterestedness. The scientist is usually concerned with the control of nature apart from his personal aggrandizement. He is perhaps not wholly "pure," but he seeks control mainly for its own sake or for the sake of furthering other scientific activity. There are practical as well as ethical reasons for this: as technology becomes more complex, for example, the scientist himself is less and less able to pursue the practical implications of his work. There is very little personal reimbursement for the most profitable ideas of modern science. As a result, a new idea may yield immediate technological improvements without bringing the scientist under suspicion of plotting a personal coup. But social technology has not yet reached this stage. A disinterested consideration of cultural practices from which suggestions for improvement may emerge is still often regarded as impossible. This is the price we pay for the fact that men (1) have so often improved their control of other men for purposes of exploitation, (2) have had to bolster their social practices with spurious justifications and (3) have so seldom shared the attitudes of the basic scientist.

Most people would subscribe to the proposition that there is no value judgment involved in deciding how to build an atomic bomb, but would reject the proposition that there is none involved in deciding to build one. The most significant difference here may be that the scientific practices which guide the designer of the bomb are clear, while those which guide the designer of the culture which builds a bomb are not. We cannot predict the success or failure of a cultural invention with the same accuracy as we do that of a physical invention. It is for this reason that we are said to resort to value judgments in the second case. What we resort to is guessing. It is only in this sense that value judgments take up where science leaves off. When we can design small social interactions and, possibly, whole cultures with the confidence we bring to physical technology, the question of value will not be raised.

So far, men have designed their cultures largely by guesswork, including some very lucky hits; but we are not far from a stage of knowledge in which this can be changed. The change does not require that we be able to describe some distant state of mankind toward which we are moving or "deciding" to move. Early physical technology could not have foreseen the modern world, though it led to it. Progress and improvement are local changes. We better ourselves and our world as we go.

We change our cultural practices because it is in our nature as men to be reinforced in certain ways. This is not an infallible guide. It could, indeed, lead to fatal mistakes. For example, we have developed sanitation and medical science to escape from aversive events associated with illness and death, yet a new virus could conceivably arise to wipe out everyone except those to whom chronic illness and filth had granted immunity. On the present evidence, our decision in favor of sanitation and medicine seems to make for survival, but in the light of unforeseeable developments we may in time look back upon it as having had no survival value.

From time to time, men have sought to reassure themselves about the future by characterizing progress as the working out of some such principle as the general will, universal or collective reason, or the greatest good. Such a principle, if valid, would seem to guarantee an inevitable, if devious, improvement in the human condition. No such principle is clearly supported by a scientific analysis of human behavior. Yet the nature of man tells us something. Just as an ultimate genetic effect cannot be reached if immediate effects are not beneficial, so we must look only to the immediate consequences of behavior for modifications in a cultural pattern. Nevertheless, cultural inventions have created current conditions which have at least a probabilistic connection with future consequences. It is easy to say that men work for pleasure and to avoid pain, as the hedonists would have it. These are, indeed, powerful principles; but in affecting the day-to-day behavior of men, they have led to the construction of cultural devices which extend the range of both pleasure and pain almost beyond recognition. It is the same man, biologically speaking, who acts selfishly or for the good of the group, and it is the same man who, as a disinterested scientist, will make human behavior vastly more effective through cultural invention.

REFERENCES

1 Skinner, B. F., *Verbal Behavior*. New York: Appleton-Century-Crofts, Inc., 1957.
2 Skinner, B. F., *Science and Human Behavior*. New York: Macmillan Company, 1953.

HENRY A. MURRAY

Unprecedented Evolutions

IN VIEW of the nearly unanimous determination of social scientists to remain competently within the realm of their own discipline—keeping away from the sea of troubles in which all of us, wittingly or unwittingly, are floundering—I am not unwilling to accept the role of scapegoat by offering a programmatic paper in the name of the most formidable and imperative problem that man has ever faced, from the solution and repeated solution of which he will never be exempt.

In refocusing attention on this dilemma, I shall be sustained by the surmise that scientists who are trained to detach themselves at will from the events and passions of their own space-time and to review what is known of the temporal span and global scope of human history, as if seated with uncommitted curiosity on a Martian hilltop, are equipped—more as consequence of this trained capacity than of their grasp of theory—to contribute an essential attitude and dimension to the total endeavor of the makers of our foreign policy to foresee different possible consequences of present trends and devise long-range, transforming strategies.

And so, with apologies to science and its venerable valued standards, I shall conclude this preamble and turn toward the unknown future—toward two obvious and unquestionable threats and one obvious but very questionable opportunity, none of which requires knowledge of evolutionary principles to perceive.

Threats and Opportunity

Threat 1: the stage by stage expansion of the political sovereignty and ideology of the USSR and/or China, directed toward the domination of the world from Moscow or Peiping. The eyes of the United

States government have been focused steadily on this threat.

Threat 2: the initiation of a series of wars with absolute weapons, resulting in massive exterminations in both northern hemispheres (with the extinction of democracy and freedom), leading on to more extensive and disastrous incinerations—universal agony, an intolerable earth, and loss of the will to live. The eyes of the United States government have been largely diverted from this threat.

The possibility of an atomic hell is mentioned every day—blandly, casually with a shrug or quip, fatalistically, despairingly—and then dismissed for the sake of pleasantness and intestinal composure, or denied as a Christian Scientist may deny the existence of a malignant cancer. There are also people who dare to look intently at this possibility but insist that the traumatic consequences of a nuclear attack on this country would not be catastrophic—no more than 20,000,000 persons are likely to be killed, and, if proper measures are adopted, eventual recovery can be confidently predicted. But these cool calculators never seem to take account of the possession of absolute weapons (atomic, chemical, and biological) by a score of nations, and a chain of subsequent wars resulting in an insufferable environment, with that proud, self-styled paragon of animals—by then wholly impersonalized and mechanized—reduced to an ignominious, mole-like existence underground.

Opportunity: (a) the unprecedented abolition of war—impossible without (b) the unprecedented establishment of unprecedented world laws, government, and police—impossible without (c) the unprecedented widespread conversion of peoples and governments to an unprecedented vision of world unity and fellowship—impossible without (d) the composition of a book, say, a sort of philosophical history of the world with multifarious components, a love-engendered book (to supersede the hate-engendered gospel of Karl Marx), which would end by realistically directing thought and passion toward the inauguration of mutually beneficent international reciprocities and the settlement of differences without resorting to massive atrocities and murders.

Now, with the possible exception of the last proposal—a sort of cry in the wilderness for a testament of human genius at its best—none of this is new. It has been said and resaid by innumerable men entitled by their acknowledged learning, wisdom, and good-will to a considerate public hearing. What could be more rational than the argument for some form of world government—government, by definition, being the only class of institutions empowered to suppress violence? In view of the history-making behavior of contentious and ferocious social units, however, the idea is likely to be dismissed as a patently absurd hope. But then, is it any more absurd than being towed, as we are now, by a cable of primitive, blind passions nearer and nearer to the verge of a meaningless, universal holocaust? The enemy is a lunatic within our psyches, on the threshold of a tantrum and yet not insusceptible to conversion.

That the atomic age called for a radical transformation of a sufficient number of personalities to make a difference was immediately apparent to Norman Cousins, who, within a few weeks after Hiroshima, had published *Modern Man Is Obsolete*. Six months later Albert Einstein said substantially the same thing in a now famous passage (italics mine):

We can only sound the alarm, *again* and *again;* we must *never* relax our efforts to rouse in the peoples of the world, and especially in their governments, an awareness of the unprecedented disaster which they are *absolutely* certain to bring on themselves unless there is a *fundamental change in their attitude toward one another as well* as in their *concept of the future*. . . . The unleashed power of the atom has changed everything except *our ways of thinking* [*The New York Times*, 25 May 1946].

Although this passage has often been quoted, I have yet to hear of a serious attempt to define exactly what fundamental changes in attitude and in our concept of the future are required to avert catastrophe. First of all, assuredly, we need men of imagination and goodwill with enough fortitude to keep staring at the current trend of events and its inevitable outcome until they realize in what respects they—as well as all the rest of us—are no longer fitted for survival, and then, in the light of this conclusion, commence a painful process of radical self-conversion. The atom has revised the rules of evolution.

With these necessities in mind, I would suggest that another subsidiary threat is the fact that so little concentrated thought is being seriously devoted to this greatest of man's perils. How does it happen that the American people, American leaders, American intellectuals are not wrestling day and night with the heroic task of devising ways to induce the peoples of the world to abolish war and establish a sound foundation for enduring peace? In dealing with the fanaticism of Communists—to whom the bulk of the world's disturbances can certainly be attributed—shaking deterrent weapons with righteous indignation seems to be our only strategy. Are we so unconsciously convinced, perchance, of their ideology's greater holding power, so awed by it, so uncertain of our own faith, that we are incapable of thinking there are ways, therapeutic ways, of converting them to a modified version of their doctrine? Why are not all the resources of creative minds, all the resources of higher education brought to bear on the solution of this problem? Is not the prevention of national suicide the government's foremost obligation? and the perpetuation and creation of culture the first responsibility of intellectuals? Though conscious of heritage, trust, and danger, we are willing, it would seem, to be utterly ineffectual, superfluous, even frivolous. Whatever the explanation, most of us—each preoccupied with his own work—have been afflicted by catatonia of the political and ideological imagination, temporal and spatial, living in the present by animal time and hence reacting with our government to a succession of unexpected stimuli (usually emanating from our proclaimed enemies), instead of living by human time in relation to the known past and to programs for a better future.

Almost the whole of this communication—for the initially-given reasons—consists of an attempt, slight as it may prove to be, to atone in part for my recurrent infidelities to the paramount (in my scales) obligation of our time, to do this by offering a number of psychological considerations bearing on a possible program to strengthen (a) the fifty-year-old, often defeated, and currently weak and halting trend toward the establishment of institutions and of grounds for peace, and, conceivably thereby, to weaken (b) the currently strong trends of expansion of both the political sovereignty and ideology of the USSR and China, and (c) the currently strong trend of multiplying animosities and multiplying armaments for war.

Unexampled Supranational Laws and Institutions

Law and government being entirely outside my sphere of competence, I cannot pass judgment on the relative merits of this or that proposal for an institution to prevent war. I can only confess my

Unprecedented Evolutions

inability to imagine the permanent abolition of wars, in the absence of institutionalized procedures for settling disputes between infuriated states and of sufficient force to insure conformity to the decisions reached. A great deal of solid, brilliant thought has already been applied to this bi-horned problem by Grenville Clark and many others, here and abroad, and it is quite conceivable that creative legal minds from both sides of the iron curtain will eventually collaborate, as did the fathers of our nation, in bringing forth a workable constitution—the substance of the first world conscience, or superego—for a loose and flexible federation of nations dedicated to the preservation of peace. Beyond this simple avowal of faith and the skeptical avowal that this is not likely to occur without a widespread revolutionary change of attitude (to be discussed in the next section), I have but two comments to make.

The justification for war has been based either on the primitive (adolescent) worship of sheer physical power—might is right as well as glorious—or on the assumption that military power is correlated with other valued manifestations of energy and vitality, on the level, let us say, of material, social, political, moral, religious, scientific, and aesthetic well-being; and, consequently, the expansion of a society's territorial sovereignty by military force results in a corresponding expansion and multiplication of better forms and higher standards of sociocultural transactions. Since it is in these terms that progressive evolution is defined, it is in these terms that a nation which grows in strength relative to other nations is inclined to vindicate, often in the name of God or Destiny, its importunate ambition to extend its boundaries, obliterating, if necessary, less viable societies. A nation in this phase of its career, puffed up with self-esteem and arrogance, will inevitably oppose the idea of a supranational institution—a world court—for the arbitration of quarrels between sovereign states, because, like any criminal, it perceives law and justice as an impeding force in the service of the weak, and also as an intolerable block to its self-appointed evolutionary role.

Today the great powers, particularly the USSR and China, are still thinking in these terms, regardless of the fact that they are obsolescent. From now on, it is precisely the proudest and mightiest nations, the possessors of absolute weapons, that are earmarked for extermination.

HENRY A. MURRAY

Now that we realize that war is no longer the servant of evolution but is its prime enemy, we are in a better position to measure its previous incompetence as arbiter of the worth of a religion, ideology, philosophy, morality, form of government, or way of life. The traditional function of war has been to make enduring decisions, despite our recognition that physical force is utterly impertinent to questions of the relative merit or truth of different intellectual propositions, ethical codes, or patterns of behavior. Still, there is no denying that in many instances wars have served to accelerate the processes of cultural diffusion and expansion; and so, if legal decisions are to replace wars, they must not all result in the perpetuation of a static global order, but allow for gradual changes and expansions determined by relevant and rational criteria, instead of by the irrationalities of force.

Those who take part in the administration of supranational affairs will have to be extraordinary characters, especially trained and educated in several different countries, experienced, mature, and sagacious citizens of the world, as disciplined and dedicated as the best Jesuits, as trustworthy, gentle, and tolerant as the best Quakers, of irreproachable integrity, and with the cultural breadth and universality of the best anthropologists—saints, in fact, of an unprecedented type.

Synthesism

This whole paper is founded on a single questionable premonition, namely, that we are confronted by two long-range options: the virtual abolition of war or the virtual abolition of man. The first would be an unparalleled accomplishment which looks impossible today, partly because the creativity of humanists has not yet produced the specifications of an effective method. The second would also be an unparalleled accomplishment, but this one is not impossible, the creativity of scientists having already produced several perfectly effective methods. Even if we had the blueprint of a method, the first accomplishment would be far more difficult than the second. Extraordinary genius, courage, and devotion would be required to convert people in sufficient numbers to an ideal of world fellowship protected by law and government, but no genius to give the word to those who are waiting to press buttons in the name of patriotism. The abolition

Unprecedented Evolutions

of war calls for a radical transformation of modern human nature; the abolition of half the population of the earth calls for nothing but submission to the existing drift of passions, national and ideological.

According to Jung (who first recognized this basic difference in people), extraverts, who are disposed to act by impulse, by suggestion, or by custom, can be taught to behave differently only by repeated failures. They must act, err, suffer the unhappy consequences, and only then, if painfully impressed, will they revise the values by which they have been living. Introverts, on the other hand, are more apt and able to revise their values on the basis of trial-and-error experiments carried on in their imaginations. Through such anticipations, some of them may become allies of the future, though often alien to the apparent practicalities of the day. Since ours is an age of rampant extraversion—the dominance in the United States of thousands of uncomplicated extraverts and in the USSR of extraverts combined with a hard core of extraverted introverts—one might surmise that only after a thorough catharsis of aggression and after experiencing the hell of nuclear explosions, will men arrive at the unambiguous conviction that modern war is of all evils the most abominable, diabolical, and lunatic, and that the continuing prospect of other, more devastating wars cancels, definitely and forever, the hope of any life that is worth anybody's living.

There are some people, in other words, with sympathies circumscribed by a deficiency of imagination or by egocentrism, who have to feel the agony in their own bodies, realize that they themselves are crippled and disfigured for the rest of their days, before they will cry out against the enormity of war, and against those who are in any way responsible, and finally against themselves as sharers of this responsibility. But fortunately there are people of another grain, whose vivid images of horrors to be endured a hundred years from now are enough to motivate persistent efforts to abolish war, or to cope with the conditions, the grievances and anxieties, the misconceptions and fanaticisms that lead to war. The question is, how soon will the forces of imagination bring enough influential people on both sides of the now permeable curtain to the point of certainty that world concord is a "must"? Can this point of enlightenment be reached without the profound anguish of a genocidal global war? Or will it take one, two, three, or four wars to drive mankind to sanity? One guess—this side of utter pessimism—would be that before this earth is wholly rid of our ferocious race, a saving remnant will bring forth a new religion celebrating spiritual unity amid diversity, with a covenant of sacred pledges, disciplines, and rituals conducive to the perpetuation of peace.

It is strange that the actual present prospect for human life on this planet—a broad road to an atomic hell or a narrow path to a federation of humane societies—corresponds in the abstract to yesterday's extravagant Christian formula for inducing individuals to repent (through dread of everlasting torture) and to live virtuously (in hopes of an eternity of bliss). At first blush it might seem that the task of the apostles and priests of Christianity was more formidable than ours, since they had to convince men and women of the reality of two contrasting after-lives, unaided by the slightest evidence of the kind required by those who hold that "seeing is believing." In the absence, say, of irrefutable exhibits and moving pictures, such as those of Hiroshima, they had only invented images to rely on.

But in several other respects those who undertake today's great task of transformation are at a decided disadvantage: (a) they cannot seriously claim (without being committed to an asylum) that they are the chosen spokesmen of God's revealed purpose; (b) their advocated course of action is not appealing to self-interest, since the goal (world peace) is not within the reach of the private faith and works of single individuals; and, if finally attained by an immense collective effort, its benefits are more likely to be enjoyed by others (posterity) than by those who toil and suffer for it now; (c) since the goal (an institution to prevent war) can be established and maintained only by the rulers of the various sovereign states, it is, above all, these (often insensitive, aggressive, and myopic) rulers of these traditionally vain, amoral social units who constitute the ultimate target of the transformation process; (d) the realization of the goal will depend on the determination, wisdom, patience, and exertions of all the major powers—which introduces a radical psychological difference, since no single nation, inflated with self-esteem, can claim the credit: every power must be prepared to share the glory of this superlative achievement; and (e) as yet, no specifically inviting images of realizable rewards—convincing illustrations of mutually advantageous and enjoyable reciprocations between peo-

Unprecedented Evolutions

ples of different nations—have been proffered. For the chief motivators of constructive efforts, therefore, one is left with threats of punishment (atomic hell-fire), which, according to contemporary theory of learning, are less effective in the long run than promises of reward.

Basic to several of these impediments is the hereditary egotism and pride of nations, especially of the big nations—that is to say, the compacted egotism and pride of the majority of their members, greedy for material possessions, power, and prestige, who, in dread of having to part with any of these sources of satisfaction or of having to reduce their hopes of more, are stoutly opposed to having their nation share even a small portion of its precious sovereignty with other nations. What can be done to override these granite blocks of national avarice and vanity—condemned in other peoples but applauded in ourselves—in order that we all may become worthy of survival in a potentially explosive world? If our present peril can be attributed in the last analysis to the dispositions that sustain the ideology of nationalism, must not our hope of deliverance be attached to the creation, acceptance, and efficacious operation of a transforming supranational ideology, an unprecedented and essentially religious vision, or ideal, for the ultimate concern of all mankind?

That an unexampled supranational ideal—a complete refutation of the glorified egotism and material grandeur of monarchs, states, and individuals—can emerge, make headway against relentless persecutions as well as against all the judgments and predictions of learned rationality, and finally triumph by officially converting the greater and lesser rulers of the entire known world—that this can happen is evidenced by the early history of Christianity. In those first centuries, the prime task was that of radically transmuting the instinctive tendencies of individuals, of turning egotism into altruism, pride and arrogance into humility and submission, wrath and brutality into charity and pity, sadism into masochism, avarice into generosity, and lust into chastity. On the basis of what historical knowledge, at the time of Augustus, could a statesman or some hypothetical social scientist have foreseen that the proudful *thesis* of the almighty Roman Empire, represented by the superordination of that common, all-too-human trinity of values—material possessions, power and prestige—and the deification of the emperor, would, to a marked extent, eventually succumb to its exact *antithesis*—meekness coming out of Palestine?

Much of the moral activity of Western minds subsequent to the establishment of Christianity can be represented as a long succession of attempts—desperate, futile, arbitrary, casuistic, contentious, bloody, irrational, delusional, neurotic, hypocritical, spurious, superficial, or cynical, continued or abandoned—to resolve obsessional conflicts between these two irreconcilable sets of values. Egotism or altruism? Vengeance or forgiveness? Defiance or compliance? Wealth or poverty? Hedonism or asceticism? Sexuality or chastity? More bitter were the conflicts over inconsequential points of faith, and, after centuries of meaningless disputations, of inquisitions and atrocious tortures, of savage religious wars and massacres, of ruthless suppressions of thought and speech, the Western mind woke up as from a nightmare, rubbed its eyes and, freed of ancient phantoms, enjoyed for the first time lucid impressions of the world around it, as had the Greeks two thousand years before. But no substitute for Christian morality was forthcoming, and now, except for a stout strand of humanitarianism, a detestation of brutality, and episodic acts of charity, we are returning, as many students of history have noted, to the chaotic state of morals and morale that prevailed in the Roman Empire just previous to its decline and fall. We are in the throes, it is generally agreed, of a period of transition, searching for a resolving symbol, or ideal. What could it be?

My own unhesitating answer, the only basic, positive proposal in this paper—obvious as the earth and yet scarcely communicable in any words available to me—consists of the multifarious phenomena included in a concept which, with serious misgivings, I shall term *synthesism*. *Synthesism* means an evaluative stress—at a certain stage of development, the greatest evaluative stress—on the production and continuation of a synthesis (combination, creation, integration, union, federation, procession of developing reciprocities or transactions) particularly of opposites (positive and negative, male and female, contrary or antagonistic entities, groups or principles).

Though applicable to the integration of *numerous* entities, synthesism can be more easily defined when it refers to the formation and perpetuation of a *dyadic system,* that is, a unity of *two* interacting and in some ways antithetical components. Novel chemical and genetical combinations, as postulated by the theory of creative, or emergent, evolution, constitute low-level analogies of the phenomena in question; but the simplest, valued human paradigm is provided by sexual conjugation when mutually and synchronously enjoyed by both participating members of an enduring affectional relationship, or dyad. Synthesism stresses the value of the dyad as such, over and above the value of the two component individuals, and judges the worth of the system in terms of such criteria as intensity, depth, scope, freedom of expression, flexibility, variation, exposure and settlement of conflicts, creative transformation, stability, and duration. A friendship that is markedly characterized by these qualities would constitute another example of dyadic synthesism on the personal level. And then, to jump to two levels of the broadest compass, we have dyadic representational synthesism illustrated by the integration of two, in some ways antithetical religions, ideologies, philosophies, scientific theories, or art forms, and dyadic societal synthesism, illustrated by the enduring friendship or federation of two previously antagonistic nations.

The word "synthesis" was derived from chemistry, but also with a new meaning, from the thesis, antithesis, and synthesis of Hegel's metaphysics. As suggested above, the egotism (pride, vainglorious ambition, competitive greed, military force and crimes) of nations could be termed the thesis, a thesis which from here on can lead only to catastrophe and hence calls for some controlling principle or reaction formation, such as Christian altruism, the antithesis of egotism. But Christianity, being concerned with the morality of single individuals, never seriously advocated altruism as an ideal for sovereign states. In Whitehead's words: "As society is now constituted a literal adherence to the moral precepts scattered throughout the Gospels would mean sudden death." Something which does not lead to catastrophe or sudden death, however, is the synthesis of this thesis of national egotism and this antithesis of national altruism, namely, international friendship, affiliation, or dyadism (each party both giving and receiving), esteemed in its own right over and above the self-estimation of each reciprocating nation. A comparable dichotomy of values on the individual level is illustrated by the personality of Jesus, whose optimally egotistical thesis (the claim that he was not just another enlightened Hebrew teacher in the prophetic tradition, but the Son of God and possibly, in a spiritual sense, the King of the Jews) was balanced, in a health-giving, fruitful way, by its extreme antithesis, a life devoted to relieving physical and spiritual distress in others. But no synthesis occurred, no entangling and exacting, joyous and vexatious commitment to a marriage or even to a friendship of *equals*—an avoidance which, with ultimately fatal consequences, excluded a good half of human life—body, woman, Eros, creativity—from the nuclear mythology of Christianity.

The experience of fellowship, in-group fellowship, is as old as society, and published expositions of the virtues of male comradeship as old as Aristotle's *Ethics,* and celebrations of courtly, or romantic, love as old as Dante's *Vita Nuova* and the legend of Tristan and Isolde. We, being familiar for so long with these traditional phenomena and their designating dictionary terms (e.g., fellowship, love, etc.), can easily overlook the crucial difference between these conceptions and the one I am attempting to set forth. This is the chief reason for the introduction of a new word, "synthesism."

Synthesism—or *dyadic synthesism*—first of all, calls for the elevation of the hardly utterable, shared values of participation in the creation and development of better forms and qualities of relationship (continuity of union, of mutual affection and respect, amid diversity of patterns of interaction) from a subordinate to a superordinate position, that is to say, the experience and fruits of affectional reciprocations, interpersonal and international, would be more highly prized than personal and national superiority and aggrandizement. To appreciate the emotional revolution involved in this transposition of values, we have only to remind ourselves that all formerly venerated models of excellence or greatness have been glorifications of a single person, a single group or nation, a single theory or religion. The monadic myth of the hero, or the monadic myth of the society with a mission, appointed by God or Destiny, is the prototype of all impelling visions—masculine visions—of the past. What chance is there of moving it to second place? of substituting a masculine-feminine dyadic myth? On the personal level, this might be effected with greater ease today than was the revolutionary change during the first centuries A.D. from the monadic Caesar myth (the deification of force and grandeur) to the monadic Christ myth (the deification of compassion and blessedness). But the wrench should not be underestimated: some time must be allowed for the reduction of the prevalent, cool monadic myth of conspicuous success (say, egocentric pride in one's heralded achievements) and the elevation of the passionate dyadic myth of mutually creative love (say, shared pride in the qualities of a periodically recreated marriage). As many social scientists have noted, there has been a decided trend in this direction since the end of World War II, especially among the young.

The second point is that we are abysmally ignorant today about the potentialities for joy and growth that reside in marriage, the potentialities of a different sort and of less degree that reside in friendship, and, of course, still more ignorant of the potentialities that reside in international reciprocities. Certainly we must acquire, interpret, and transmit far more explicit knowledge and wisdom than is now at hand about various types of dyads and the necessary stages in their development, about impeding and facilitating, frustrating and fructifying modes of interaction, about varieties of experienced penalties and rewards, about ways of settling critical disputes, and about temporal changes in the characters of the two participants. Without this knowledge we can hardly say anything definite or profound about synthesism. The bulk of the scientific literature on marriage—there is very little on friendship—is either statistical and superficial or is chiefly concerned with the determinants of unhappiness, discord and divorce. One reason for the deficiency of knowledge and wisdom about better ways and directions of self-realization and development *within* an evolving dyad is that these have yet to be imagined, tried, experienced, evaluated, and suitably represented.

In other words, *dyadism* (which stresses shared experiences of cycles of relational creation, conservation, and re-creation) calls for the focusing of thought and representational abilities on the progression of these processes. Heretofore, the creative energies of men have been almost invariably directed toward other goals—in some cases toward the invention of weapons to destroy human relationships—rarely toward the invention of multifarious ways of enriching, deepening, and sustaining them most happily. As suggested earlier, the gravest flaw in Christianity, in contrast to certain Hindu sects, is the absence of an inviting and illuminating myth, a set of guiding symbolic models, to represent the evolution of a passionate erotic dyad, and so today the young who enter marriage in the hope of fulfilling their deepest needs and highest aspirations have no possibility of succeeding unless they are disposed and able to create as they proceed a mythology consonant with their experiences.

If some serious and gifted modern writers and painters could arrive at the realization that the meaning of nature, the meaning of human history, the meaning of experience are never *given* by nature, history, or experience, but are always created by man's imagination (in sensible or conceptual language), they might be cured of their addiction to excremental representations of chaos, formlessness, meaningless violence and absurdity, combined with their feelings of disgust, impotence, and alienation. Why do they not choose to play some part in the creation and symbolic representation of new forms of meaning, instead of wasting their time deploring the demise of Christianity or the resulting hollowness of modern men?

My third and final point is that satisfying creative reciprocations is the key concept, or *sine qua non*, of synthesism, with cooperations and goodnatured competitions and oppositions included as subsidiary values. There has always been an abundance of in-group cooperation in the world, no society could exist without it, and today it is more widespread and insistent than it ever has been, it is being enforced in the Soviet Union on a total scale, and enforced in the United States and other countries on an institutional scale (e.g., military and industrial "organization men"). Necessary as they are and enjoyable as they certainly can be, cooperations do not constitute the core of synthesism, being, as they are in war, so often directed towards the exact opposite—destruction rather than creation.

If successful, one world would constitute a social condition unprecedented on this planet: a single society of diverse nations, united by bonds of affiliation and reciprocation, which were formed and which maintained their holding power without the compelling stimulus of a threatening outside competitor or enemy. History informs us that intrasocietal solidarity is almost invariably increased by the necessity to cooperate, felt whenever a whole society is challenged by the awareness of danger to its integrity or by the exciting prospect of conquest, of gaining more territory, wealth, power, and glory, in which all might have a share. The regularity of this phenomenon has entitled it to the status of a political axiom, so generally accepted that a ruler who has become the focus of his people's hate will search for some pretext, or invent one, for declaring war in order to set up another target for the catharsis of their hostility. But cooperations of this sort, within groups and between groups (alliances, axes, blocs) are likely to fall apart when the need for them ceases to be urgent. Consequently, although many forms of cooperation—scientific cooperation, cooperation to solve the population problem, the hunger problem, the cancer problem, and so forth—are effective and fruitful ways of forming and solidifying bonds of affiliation between peoples, patterns of reciprocation (which may, of course, occur in the course of a collaborative enterprise) are nevertheless more basic to the achievement of a synthesis.

Many forms of valued reciprocations (transmissions and receptions)—of affection and respect, appreciation and encouragement, material goods and services, information and special knowledge (theoretical and technical), interpretations and evaluations, new ideas and programs, entertainment, dramatic, gymnastic, and musical performances, and so forth—have occurred—with certain rigid limits of discourse, to be sure, and yet sometimes to a notable degree—between Russians and Americans in their meetings here and there outside the domain of political strife and agitation. The extraordinary degree of spontaneous enthusiasm and good-will (especially on the part of the Russians) repeatedly displayed on these occasions—at conferences, recitals, athletic contests, and scores of less formal meetings—shows that the *peoples* of these two lands have much in common temperamentally, they are a virtual well of potentialities for fruitful affiliations, and, when separated from their governments, they have not the slightest inclination to exterminate each other. But all this leaves out imbedded ideological and moral differences, and, until synthesism is successful on this superordinate level, no dependable future can be anticipated.

Let this suffice as a summary account of synthesism operating at different levels. I have illustrated the concept by more references to dyadic interpersonal than to polyadic international transactions, because the former are simpler, closer to everyday experience, already partly achieved by some people, and more intimately related both to primitive human nature and to Christianity (singularly unsuccessful in preventing wars). Furthermore, synthesistic experiences on the personal level would ordinarily be antecedent to participations on the higher levels.

To hold that synthesism provides the basis for a positive and creative morality (as a necessary supplement to the negative, prohibitive, static morality we have inherited) and that it constitutes the fundamental answer to the world's dilemmas at all levels of violent opposition sounds extremely simplistic, sentimental, and utopian. Well, it *is* simple in the sense that it depends on one thing: a radical conversion of heart and mind marked by the explicit adoption of a new, positive, superordinate direction of creative effort by two or more parties. Once this emotional shift has taken place, once the participants have veritably pledged themselves to persist in a united effort to resolve their differences and to surmount conflicts by devising better, more varied and more rewarding interaction, then *everything* is simplified, the accomplishment of all tasks greatly facilitated, and, if the compact holds, no problems, however complicated, should permanently resist solution. But, of course, I am not suggesting for a moment that this transformation—the way out of the obsolescent states of rampant egotism and rampant nationalism —is simple in the sense that it is easy, or even that it is likely to occur soon and on a large enough scale to make a vital difference.

As indicated earlier, I do not belive that any such a revolutionary conversion of human nature is conceivable without the composition of a book, or book of books, the skeleton of which might well consist of a condensed metahistory of mankind (say, encompassing and correcting Marx and Toynbee), brought to life by myths, legends, stories, anecdotes, parables, aphorisms, and poetic passages illustrative of human relations and their vicissitudes, of war and peace, of feuds and friendships, of conflicts and their resolutions, a book which would end by setting forth in some detail, with pithy and telling examples, the principles of synthesism, so briefly and inadequately outlined in this section. Such a book would be designed to provide the needed philosophical and moral basis for a creative foreign policy, as well as for the dispositional education and self-development of individuals. I believe that we can find scattered through the various literatures of the Orient and Occident many suitable passages and chapters for this book of books, some written centuries ago and some quite recently. And, just as the body of scholars appointed by King James produced within seven years the authorized version of

Unprecedented Evolutions

the Bible, so today, granted some genius in the marrow, a similar body might compose, partly by writing and partly by judicious cullings from other authors, a preliminary version of such a transforming book to guide the peoples of the world toward unity and peace.

Since the present international furore has its origin in the frantic, fanatical mystique of Communism—initiated and sustained by the testament of Marx—and since this programmatic doctrine—having proved to have unprecedented power to rejuvenate, energize, solidify, and orient underprivileged societies *en masse*—evidently has some future on this earth, in part or as a whole, and since we cannot possibly come to peaceful terms with its existing form and current modes of operation, and cannot possibly exterminate it totally with bombs (threats serve only to increase its furious momentum), only one possibility remains, namely, to transform it: to alter, to some extent, the nature of its objectives, to reduce the magnitude of its aim, and to moderate the violence and importunity of its propagating methods. This may be less difficult than we imagine, since the doctrine of Marx, valid as it is in several of its basic tenets, is full of flaws and errors: for example, its metaphysics of human nature, its restricted economic focus, its hate-engendered delusions of monstrous robber barons, its absurd visions of an utopian end state, and, naturally, its omission of atomic energy and the possibility of an utter defeat of all efforts to improve society. Anyhow, the book I have in mind, written with a clear understanding and appreciation of the instigations, doctrines, and purposes of Communists, would address itself to the task of their conversion to a synthesis of value orientations, and to a policy of gradualism. Like children, we are prone to feel that we shall win this serious and portentous war for men's minds—not with better ideas, not with a deeper and better book that comes to grips with the basic realities of human nature and defines a realizable vision for mankind—but by better spacemanship and missiles, better spies in the sky, and better recoveries of better astronautic chimpanzees.

Toward a Re-Orientation of our Foreign Policy

In the United States pretty nearly everybody is allowed to speak his piece: catharsis reduces tension, and who can say in advance from what head or what class of people a usable idea will emanate.

HENRY A. MURRAY

Whether the occasional advent of a really good idea from an unlikely source is sufficient remuneration for the enormous amount of time spent attending to a perpetual parade of poor ones is a moot question. But whatever the cost, freedom of speech is guaranteed to all of us by an enlightened and resilient government—not a right to be taken for granted or abused, but (in the history of nations) a rare privilege, especially prized by those who have aspirations for their country and feel involved somehow in a joint push toward their attainment. The high probability that speaking in public will merely add to the circulating sum of poor ideas is certainly a deterrent. But who knows? Who can foresee the interplay of chance and circumstance, good-will and bad-will, sanity and madness?

The few things I have to say in this last section are based on the belief that the abolition of war would seriously disturb, strain and dislocate large sectors of the economic, social, and political structure of this and other countries, thereby giving rise to a multiplicity of formidable problems of reconstruction, but that these could eventually be solved, provided the determination to abolish war was strong and resolute. Nothing can be done without a firm, widespread, sovereign desire for a world of friendly nations. With this in mind I shall restrict myself to several unsolicited suggestions as to a few fairly obvious ways of reducing tensions between the great, war-ready, clenched antagonists of our time, as a necessary prelude to the reception of a world view and plan which will be set forth in broad outlines at some timely moment in the future, its details to be worked out later, step by step.

My assumption is that we shall maintain our military striking power until a suitable schedule of disarmament and a dependable method of inspection have been definitely established, and that in the meanwhile we shall approach the Russians and the Chinese with a bomb in our retracted left hand (not brandishing it in the world's face) and an olive branch in our extended right, ready to seize any opportunity to break through whatever tactical postures and poses a dictator may present and get at the heart of his version of the crucial issues. My suggestions are as follows:

1. *Change from a negative to a positive orientation.* Announce at a strategic moment that our objective is the abolition of war, or world concord, with the settlement of international conflicts by the

institution of a new form of world law and government, and a world police force as deterrent against war. Set forth this vision as the only alternative to world domination by one state. What country wants to operate under the heel of the USSR or China? Point out that the adoption of Communism by every state in the world—many of them in possession of atomic weapons—would be no guarantee of peace.

2. So far as the resources of courage and persuasion will permit, encourage the latent potentialities of heart, intellect, and art in the American people to become involved in the process of converting ourselves and other peoples to the paramount ideal of world fellowship and to the ways and disciplines of attaining, enjoying, and sustaining it. The announcement of this aim should stimulate, in due course, the gradual creation of an imaginative symbolism, or mythology, leading to new art forms, to international and interpersonal rituals and festivals, which would restore to human life the missing depth dimension. The ideology of synthesism is based on the conviction that most people in their innermost selves prefer harmony to discord, affection to disaffection, peace to war, life to death, and, in view of these realizable possibilities, can be educated from birth to moderate their excessive aspirations for profit, property, power, and prestige.

3. Stop defining the current situation as a religious or ideological war. So far as possible, stop using the word "Communism": stop proclaiming that our policy is to "fight Communism" wherever it exists:

(a) Because "Communism" is a word with a religious significance and potency, symbol of a mystique, to which millions of people are devoted as their tested remedy of ancient ills. Expressions of implacable hatred of Communism can only serve to increase the fanatical energy and drive—and hence the achievements—of its supporters. So long as we provide veritable ground for the magnification of the image of our nation as the dragon enemy of their whole system, the morale and present degree of productivity of the peoples of the USSR and China will certainly persist or mount. Under ordinary circumstances, the basic problem of a socialist economy is how to maintain the motivation of the workers, but if fate happens to present them with the challenge of a menacing competitor or opponent, the problem ceases to exist. Moral: lessen the menace.

(b) Because by announcing that we are irrevocably opposed to Communism, we strengthen the bonds of affiliation between Communist countries. They will stand solidly together so long as they have a powerful and belligerent common enemy. But if we say—what is closer to the truth—that we are opposed to any single state that purposes to dominate the world, our position will accord with the sentiments of other countries, even of the satellite countries (such as Jugoslavia, Hungary, and Poland), and conceivably of the Soviet Union itself *vis-à-vis* the inflated ambitions of China. In any case, we shall be weakening the links that now unite the Communist bloc.

(c) Because if we hold that it is Communism we are fighting, we shall automatically become the enemy of every country that is converted to Communism, being the presupposed cause of all the grievances of its masses and hence the target of their vindictiveness.

(d) Because—though sturdily opposed to Communism in our own country and fully justified in suppressing the virulent activities of its agents here—it is against our avowed principles to dictate to other states what religion or what form of political or economic system they should have. Today, Moslems, Buddhists, and adherents of numerous other religious faiths, as well as virtual dictatorships and socialist states, are all included as potential allies among the recipients of our aid.

4. Break up the ideological dichotomy of Communism and anti-Communism into its component opposites, say:

(a) state capitalism (or economic socialism) *vs.* private capitalism,

(b) autocracy and total control *vs.* democracy and partial control,

(c) naturalism *vs.* supernaturalism,

(d) hypertrophy of ambition for world power *vs.* moderation or suppression of this ambition, and

(e) the importunate conversion of other peoples by covert penetration, deceit, and violence *vs.* their gradual conversion, if any is attempted, by forthright persuasion and demonstration.

The effect of doing this should be to deprive the word "Communism" of part of its unholy power to arouse bellicose emotions—aggressive enthusiasm, on the one hand, and aggressive detestation, on the other—and should allow us to define the real enemy of our nation and of all other nations, namely the exorbitant ambition for world power and the devilish means employed in furthering this ambition by the USSR and China.

5. Define the irrevocable position of this nation as grounded in democracy with its partial controls and partial freedoms. Do not use "freedom" indiscriminately as an absolute, but couple it with "responsibility," and specify the kinds and degrees of freedom which we as a people have enjoyed and with what consequences (some of which, we must admit, have been beneficent neither to the country as a whole nor to the individuals [e.g., juvenile delinquents] who abused them). Acknowledge that up to now only a small proportion of the world's nations have succeeded in maintaining a democratic form of government and perhaps de Tocqueville was correct in concluding that the Anglo-Saxon temperament—cool, patient, and empirical, in contrast, say, to Latin, Germanic, and Slavic temperaments—was peculiarly fitted for the operation of a system of this nature. It seems that even highly civilized, individualistic France requires at this late date a fairly large degree of autocracy at the seat of government. Consequently, the only realistic position for us to take is that democracy is definitely good for us, after generations, first of British and then of American, political experience and discipline, and evidently it is equally good for Switzerland, the Scandinavian countries, and certain others, in several of which it has proved itself compatible with a considerable degree of economic socialism—but that we have no intention of pressuring any other country to adopt it, particularly if they are not sufficiently prepared, or of condemning those who fail to make it work.

6. If numerous knowledgeable American observers are correct in their conclusion that state capitalism (coupled with some measure of autocracy) is the only feasible economic structure for those more or less primitive societies (e.g., in Indonesia and in Africa) in which no private capital exists, we are faced with the probability, if not the certainty, that sooner or later *this* will be the structure of their adoption (with the confiscation, as in Cuba, of all property owned by foreigners). If this is true, we can either gain their friendship by helping them inaugurate the forms of industrialization that are most suitable to their needs, or turn them into enemies by letting China or the Soviet Union bring them round to it in *their* way. Also, it

certainly looks as if the future will bring forth a succession of economic and social revolutions throughout Latin America, the only question being whether they will take place in terms of Communist or other dictatorships, or in terms of democracy and liberty. It is probably within our power to decide.

7. Assured that democracy, with a wide range of personal freedoms, is the ideal political structure, and, therefore, that all countries will desire and achieve it in the near or very distant future—if there is to be one for humanity—assured of this, we should applaud every step toward a more representative form of government or a greater degree of individual freedom that is taken by other peoples (instead of criticizing them, as we do now, for their failure to meet our standards). I happen to believe that, despite its present tyranny, Russian Communism constitutes (when compared to the previous Czarist tyranny) an advance in the direction of democracy, largely because of the enormous expansion of education and of the practice (within set limits) of discussion and decision-making on the lower levels of the social hierarchy.

8. Besides commenting favorably on every extension of the range of freedom in the USSR (conspicuous since Stalin), we should congratulate them on their truly astonishing achievements in several different fields. Respectful of our judgments, they are as hungry for praise as we were for fifty years or more after our own glorified Revolution (though—boasting as loudly as we did—we rarely succeeded with the British). Instead of complacently focusing our derision on their meagre supplies of purchasable utilities (as if we, so addicted to our comforts, were unable to realize that most of them were proud to go without such things for the sake of an envisaged future), we should welcome every increase in the flow of their consumer goods, because the possession of these will make them more like us—more satisfied with their present lot, more bourgeois than they are already, nearer to corruption.

9. Instead of exaggerating differences between the Soviet system and our own, we should stress in what respects and to what degree we are alike and every year becoming more alike: for example, class distinctions have been increasing in the Soviet Union, decreasing here; and the unavoidable involvement of the United States Federal government in the support or partial control of numerous, formerly

independent systems of activity—industry, business, agriculture, communications, transportation, health, education, scientific research, social security—constitutes a decided trend toward socialism, as does the differential taxation of incomes. Also, for some years now, athletics (among other things) in the USSR have adopted numerous American forms and been pursued with the same "sporting spirit" that we inherited from England.

10. Establish an anthropological, social science institute, as an adjunct of the State Department, devoted to the collection, interpretation, codification, and transmission of knowledge about the peoples of other countries, especially Middle Eastern, African, Indonesian, and South American: (1) to serve as a center, clearing-house, and coordinator of numberless independent enterprises of this class that are being carried on in the field and at home, under the sponsorship, most commonly, of a university; (2) to serve as a center of instruction and preparation for suitable young men—chosen after six months of military training—who, after learning the language and customs of the people of a particular country, would live there as *they* do for a period of years, fulfilling one helpful function or another. This could be facilitated by bringing to the institute, as informants and as teachers of the given language, appropriate representatives of each country who would live with the American students in dormitories during their period of residence. This plan comes out of information I have received to the effect that hundreds of Americans involved in various foreign aid programs have not been sufficiently prepared linguistically, ideologically, or psychologically, to make the most of the opportunities that are offered them: to respect and win the respect of those with whom they live, to influence them in beneficial ways.

Naturally, like everybody else, I have plenty of other possibilities in mind: on the ideological level (how to learn to convert Communists), on the diplomatic level (how to talk to the Russians, analytically and candidly, with cards on the table, as they do in Thucydides' history), on the psychological level (intimate psychoanalyses of each dictator as background illumination in planning conferences), on the home-front (changes in certain American modes of life, in the treatment of minorities, in the way that nearly everything of worth and dignity is commercialized and vulgarized *ad nauseam*), and so on. But this is enough, more than enough for one paper.

Cultural Evolution as Viewed by Psychologists

THE COMMENTATORS on this topic are: Henry A. Murray, Professor of Clinical Psychology, Harvard University; B. F. Skinner, Professor of Psychology, Harvard University; Abraham H. Maslow, Professor of Psychology, Brandeis University; Carl R. Rogers, Professor of Psychology, University of Wisconsin; Lawrence K. Frank, Visiting Professor of Sociology, Brandeis University; Anatol Rapoport, Mental Health Research Institute, the University of Michigan; and Hallock Hoffman, Center for the Study of Democratic Institutions, the Fund for the Republic.

H. A. Murray: Skinner, I wonder whether you wouldn't be willing to illustrate by some specifications as to who rewards whom, for what, at what time, under what conditions, at what place, for what end—and what kind of personality is to be produced.

B. F. Skinner: It seems to me useful to compare behavioral technology with the much older physical technology. If governments, religions, and similar institutions are likely to use behavioral techniques based on laboratory research, then what has happened in physical technology may show us what is going to happen in behavior.

A man invents something—say, he discovers a new way of making a wheel. He makes one wheel and it falls apart; he makes another and it stays together. The good wheel is reinforcing for many reasons, and the right way to make a wheel is born. We are not concerned here with the natural history of the wheel or of invention in general, but we are concerned with the inventor as a man. I do not believe you would raise questions of value judgment in accounting for the invention of the wheel. The wheel defines its own goodness; it rolls, it carries weight without falling apart, and so on. Similar processes would lead to the invention of other parts of wheeled vehicles—a wagon, a carriage, a train. Each of these would work or would not work in obvious ways to reinforce the behavior of the inventor selectively. Now consider a social invention. Someone discovers an effective way of talking to an employee or a colleague or a foreign ambassador or a child to obtain a particular result. These are small discoveries, each of which is reinforced fairly immediately. But to talk about more remote effects—say, the goals of democracy—would be equivalent to asking our inventor of the wheel to predict the future of vehicular transportation.

Scientific knowledge may make a difference, however. Science permits you to look further ahead, as it did in early stages of the Manhattan Project. But this is only because science has made it possible to predict a more distant future and hence to bring remote consequences to bear on current activities.

H. A. Murray: Skinner's statement is very clear in the realm of the evolution of technology—let's say, if you want to use that term, "to control people"—but for anyone in the role of a controller, like a parent, the question would arise, "Am I going to make a child that rolls or one that flies, and also, in what direction is he going to roll, and what direction is he going to fly?" I would say that these goals are the first questions, and that you seek the methods for attaining them later. If someone gives you methods, perhaps you can apply them, but perhaps they do not apply to the end you want.

B. F. Skinner: But what if they do? If you are a parent and your child is annoying you, and you figure out a way of handling him so that he stops annoying you, that is a small technological improvement. If it works well, the authors of books on child care may hear of it, and before long millions of people may be following your example. There has been some kind of change in the culture. But Dr. Murray, as a psychologist, might object. He might argue that children are thus made too happy for their own good. He would be speaking as a scientist who has discovered through his research that some techniques of behavior may have remote consequences in the light of which we may judge them to be objectionable. This is the point I am making—that as soon as scientists demonstrate connections with the future, they modify their practices. This is certainly what has happened in physical technology, and it is, I believe, happening in behavioral technology today. The fact that you can raise the question of the relevance of practices in the care of children to the happiness of adults testifies to the importance of a science which establishes relations between the behavior of an adult and what happened in his childhood.

A. H. Maslow: I have a feeling that concealed in Skinner's presentation and basic to it is a particular conception of control which is only part of the picture. I would like to present a more inclusive approach, showing that there is an alternative to Skinner's, that his is not the only way to handle the fundamental issue of these discussions: what shall we or can we do with our fate and our future?

I think of this as an integrating proposal, and it should help in the debate between Skinner and Rogers. Ultimately, what I have to say comes down to a philosophy of science, of its process and its method, in the light of which the Skinner conception is seen as true enough, but too narrow, not inclusive enough. That is, he speaks consistently of control as if it were only an active and interfering force. This is too one-sided. I can illustrate Skinner's approach to science by listing the key words which occur again and again in his writing: lawfulness, orderliness, control, prediction, consequences, results, exactness, rigor, certainty, manipulation, shaping, molding, teaching (in the active sense).

To this overactive and interfering conception of science, I would like to contrast a very different philosophy of science, which does not really have a name yet. Perhaps I can call it for the moment, "Taoistic science," or "understanding science." Its key characteristics are receptivity to knowledge (instead of active grasping, manipulation, management), understanding as the main goal of science, rather than prediction and control (knowledge as an end in itself rather than as a means), the freer use of intuition, empathy, and identification with the object of study, a greater stress on experiential knowledge, a less pragmatic attitude.

Some examples of this approach to knowledge are "uncovering" psychotherapy (which is also a cognitive tool, like a microscope), ethnology, and ethology. These are (or should be) noninterfering, permissive, accepting, receptive to the data much of the time. This is in contrast with the experiment, designed in advance to test a hypothesis, which puts a "Yes" or "No," true-or-false question to nature, implying that you already know a great deal—just exactly *which* questions to ask.

Extremely obsessional people who must live by control, by prediction, by law, by order, by manipulation, are marked by not being *able* to let go. For instance, urination and defecation can present problems for them, problems of the voluntary giving up of control, voluntary nonvolition, so to speak. Andras Angyal has pointed out that he has never known an extremely obsessional person who could float in the water, he cannot trust the water, he has to make an effort of some sort, he has to *make* himself float, he has to strive, to float—therefore, he cannot float.

If science is primarily a tool for allaying anxiety, then the conception of active control fits in very well, and is understandable. If, however, science is also to be a path to the growth of the human being and of the human species, then we must enlarge our ideas about activity and control.

B. F. Skinner: Maslow has drawn a distinction between a predictive science and one which undertakes to control its subject matter. I envy those who can stop short at the stage of prediction in dealing with human behavior because it frees them from certain responsibilities. In psychotherapy, in particular, if you can confine yourself simply to discovering the condition of the patient and do not meddle with it, then you will not make mistakes for which you can be held accountable. However, most therapists want to do something about the conditions of their patients. I am interested in a science which is concerned with what can be done as well as with what can be known. As you know, I feel that both Maslow and Rogers are not content with mere knowing and, in fact, that they exercise some sort of control. I should like to find out more about it. I have not seen either of them in a therapeutic situation, but if I were to do so, I should watch for signs of the manipulation of variables, even though subtle, and I should expect that the effects I might observe would be explained by the processes which I deal with more explicitly elsewhere.

Carl Rogers: The paper Skinner has presented leaves me with one very strong feeling: I quite seriously raise the question whether it is not culturally time for a reformulation of our philosophy of science. It appears to me that in behavioral science, there must be room for the logical positivism, the controlled experiment, the controlled observation, which have been talked about, but there must also be room for the existing subjective person, as Kierkegaard would have called him. The reason this issue has become pressing is that, as we have moved from the study of outward events and forces to inward things, we get closer and closer to the core of what is the person, and this involves a somewhat different issue. Even the physician can consider himself as an object—he can study the infection in his hand, and regard that completely objectively, as a scientist. But when you come to the question of motivation and the other aspects that we will increasingly investigate as behavioral science becomes more subtle and more refined, I think we run the risk of really damaging, or neglecting, or doing away with one of the aspects that really makes us human.

I think that in some of the things Skinner says is a good example of some of this danger. It fascinated me that in his talking about the redesign or the reformulation of culture, he certainly avoids, just as far as he can, saying anything about values, or choice, or purpose, except as purpose may be observed simply as a nonpurposive sequence of events.

From what I understood Dr. Skinner to say, it is his understanding that, though he might have thought he chose to come to this meeting, he might have thought he had a purpose in giving this speech, this is really illusory, and that he has actually made certain marks on paper and emitted certain sounds here simply because his genetic make-up and his past environment had conditioned his behavior in such a way that it was rewarding to make these marks on paper and rewarding to emit these sounds, and that he as a person does not enter into this. In fact, if I grasp his thinking correctly, from his strictly scientific point of view, he as a person does not exist.

I am not willing to take such a dim view of Skinner or of myself or of man. I prefer to live with what seems to me to be a genuine paradox, which perhaps some day will be resolved by an overarching conception, but to which at the moment I see no resolution (and consequently I have to be willing to accept it as a paradox) that in our pursuit of science, we are fools if we do not assume that everything that occurs is a portion of a cause-and-effect sequence, and that nothing occurs outside of that. But I also feel that if we adopt that point of view in our living as human beings, in our confrontation with life, then that is death. So I raise very heatedly the question of the place of choices and values.

The second point I want to make is that there is more than one direction possible for the behavioral sciences, for the cultural evolution of man. Skinner has mentioned the possibility of control. I would like to mention at least one other possibility. The behavioral sciences could move toward the release of potentialities and capacities. I think that there is enough work already done to indicate that one can set up conditions which release behavior that is more variable, more spontaneous, more creative, and hence in its specifics, more unpredictable.

To me, it seems more in line with the basic pattern of evolution, as I understand it, that variance rather than rigidity leads to constructive evolution. What concerns me is that the more we move toward the control of men's behavior, the more I think we build into our system a rigidity that could not possibly be evolutionary in any sound sense. Actually, it is by building in rigidity that we could well bring the evolutionary process to a conclusion, while the more we aim toward the release of man's potentialities, the more, it seems to me, we shall be in the mainstream of the pattern of evolution as it seems to have operated in other realms.

B. F. Skinner: I do not feel that determinism is in any sense a threat to the individuality of man. Suppose it is true that we are all here as the result of inscrutable and inexorable forces. Each of us is, nevertheless, an absolutely unique locus through which all sorts of historical lines are passing. We have different genetic backgrounds, we have different cultural backgrounds, we have had different educations, and so on. When we confront each other as we are now doing,

COMMENTS

each constituting the immediate environment for the other, important things may happen, and they in turn may be unique. We are not determined in the sense of being fixed or static, any more than the universe is fixed or static. We all develop in various ways and will, I suppose, continue to do so.

As to value, so far as I can see, a value is simply a way of describing what is either immediately or in the long run reinforcing to man. It is no accident that we use the same word "good" for the coffee and crullers we had this morning and for the life we want to lead. Values are values because of certain properties of organisms. What has value for a horse does not necessarily have value for me. I do not relish uncooked oats and hay.

The idea of the release of potentiality, in place of control, comes close to what Dr. Maslow has said. It suggests that there are two ways of acting on the world. I thought we had agreed that there was only one. Exploding an atomic bomb is, in a sense, a release, but it exemplifies a superb control over nature. If, in working with a patient or student or friend, one arranges conditions so that he becomes more active than before and more adaptive, this is progress, but it is also control.

Of course, it is important not to freeze everything in a steady state without respect to growth and development, but there is nothing about control which implies regimentation or uniformity in that sense. You can control effectively by guaranteeing variety or, in other words, by creating "accidents." In fact, scientific method might be regarded as the deliberate construction of "accidents." Nature does not ordinarily let the scientist observe one variable under the control of another in a wide range of magnitudes. The scientist himself must vary the magnitudes. What nature has not done, he must do. He submits a system to many more conditions than are likely to occur in nature. The same thing is true of "spontaneity." You arrange conditions under which things happen, and they happen not because you force them to happen, but because you give them the opportunity. One can arrange for spontaneous combustion, not by setting something on fire, but by arranging conditions under which it catches fire. I suppose that something of this sort is true in some kinds of psychotherapy. Except for Dr. Rogers' reservations, which

I am inclined to attribute to temperamental differences between us as to the extent to which we wish to influence the lives of people, I do not think there is anything in the fundamental formulation on which we differ.

H. J. Muller: I want to say to Rogers that I believe fully in determinism, while at the same time I agree entirely with him on the need for the development of what he calls "spontaneity and creativity." There seems to me to be no fundamental contradiction between his position and Skinner's on that point, and I agree with Skinner also. However, I think there is a need for integrating these positions. This can be done, provided that we consider the values implicit in both views, even though Skinner might not wish to use the word "values" in these connections.

As I use the term, however, it only refers to what we want, and to how much more we may want this than that, or this after we get that. I do not think any of us would defend a merely drifting attitude, or one of saying, "We don't know where we're going, but are glad to be on our way," as some scientists do when they think in narrow terms—that is, there should in any case be recognized aims.

Thus, I don't think it will work simply to get better and better methods of controlling people psychologically, and to use these methods more and more, unless at the same time we increasingly clarify our ideas of what we want to do with these methods.

In fact, I do not believe there is any *ultimate* goal, because I think we shall always be in evolution. At least, I hope we always will be in evolution, and we've got to act as if we held that hope, so that we can continue to get further and further along. But that very word "further" does imply the idea of values. Skinner himself sometimes used the word "advantages," but that word also implies values or goals. So the matter can be stated in different terms, yet we cannot really judge whether or not a thing is advantageous except in relation to the values or goals sought.

Similarly, one might use the term "reinforcement" instead of "advantage," and declare that we are seeking that. I think you must ultimately come to what so many people refuse to acknowledge nowadays, namely, what John Stuart Mill called "the greatest good of the greatest number." Perhaps a better way of putting it might be "the greatest over-all happiness," or "fulfillment," or "reinforcement," or "advantage," on the part of people in general, rather than of any one individual person.

At the same time, this is not at bottom a question of the individual versus society. That is an antithesis which is very much overdone nowadays. For we do have it in our natures, both as a result of our heredity and of the bringing up accorded us by our culture, to experience reinforcement (if you wish to use this term) in consequence of the reinforcement experienced by others. Moreover, this is especially the case where we ourselves have helped to bring about the reinforcement that others have experienced. Thus, the pursuits of personal and of collective "happiness," "advantage," or "reinforcement" can be brought into harmony with one another.

There is a long period in people's lives when they are children, a long period behind that when they are embryos, and an ever so much longer period behind that again when evolution was taking place in the germ plasm from which they were derived. In the germinal and embryonic periods, and in part in that of their childhood, the processes that shaped them were not under their own conscious control at all. However, fortunately for them and for us, they were molded in such ways that at last they became possessed of intelligence and cooperative propensities. In that way they became free in the only meaningful sense of the word "free": that of being able to make intelligent knowledgeable choices of courses of action, choices directed to their own prospective reinforcement and advantage, and to the advantage of people in general. This being the case, why should it not be legitimate for us consciously to help shape even better the earlier phases of human development, those that take place before free choice can enter, in such ways that when it does it can be exerted still more effectively?

It seems to me that this aim is fundamentally the one that Skinner is pursuing. If we branded this aim as a too dangerous one, we should also, to be consistent, have to give up most of our other conscious attempts at human betterment. My own proposals for the control of the genetic background, for example, might be regarded as in the same class, so far as the question here at issue is concerned, as Skinner's proposals for "manipulating" the individual during the course of his psychological development. It is certainly not legitimate to object to such methods on the ground of their artificiality, inasmuch as it is in man's nature to resort to artifice wherever he finds it advantageous.

L. K. Frank: We have discussed at length how man has faced and coped with his life problem, and we have conceived a culture as a product of these purposive endeavors. May I suggest, however, that man created culture because he became bored with food, fighting, and copulation, and sought to make human living more meaningful and fulfilling than sheer organic existence. We may then think of culture as an artistic creation, the product of the gifted imagination of poets, artists, dramatists, prophets and the proto-scientist, who created art and religion, the beliefs and the assumptions, the expectations, the patterns of perception and techniques whereby man could transform nature, according to his hopes and fears, his beliefs and expectations, into a symbolic world for human living and purposive striving.

We may think of culture as controlling human behavior by building into the young impressionable organism these perceptions and patterns of conduct according to the symbolic world in which the child learns to live. From another angle we might consider culture as a cognitive map of the world, as expressed in the basic concepts and assumptions of culture, or we may consider it as a master code for decoding and interpreting the many messages received from the external environing world and from his own internal organic environment.

Each cultural group has selectively recognized, cultivated, and rewarded only some of man's potentialities, and it has ignored, denied, and often rigorously suppressed other potentialities in rearing the young. Thus the child is partially freed from the coercion of his own organic needs, functions, and impulses; but he pays a price for this emancipation by becoming subject to these cultural symbolic controls and to those who can and do manipulate these symbols and invoke various symbolic sanctions for ordering and controlling people.

The long accepted dichotomy of culture versus personality may be resolved if we recognize that culture is a statistical concept, emphasizing the recurrent regularities and persistent relations exhibited by members of a group, while personality is a clinical concept for the unique, identified individual who may participate in maintaining these group regularities but never fully conforms, always utilizing these cultural patterns in his idiomatic way for the goals and purposes he individually seeks to attain.

All over the world today, historically developed cultures are breaking down, losing their former integrity, as people abandon their long accepted sanctions and symbol systems. To use the old expression, the "cake of custom" has been broken; and as the "unseen hand" of tradition no longer provides guidance for individual and group living, people are becoming troubled, uncertain, and anxious, facing increasing confusion and conflict which they are unable to resolve. Increasingly, the people of the world are trying to live in two cultures, as C. P. Snow has described it, accepting modern science and technology for meeting many of their exigent needs and for furthering their new aspirations, but at the same time trying to live according to traditional ideas and beliefs that are increasingly incongruous and in conflict with their technologically patterned activities.

We can say, therefore, that the basic problem facing the peoples of the world today—and this applies to us as well—is to renew their culture, utilizing our growing scientific and artistic resources not only to reconstruct social order but also to pattern the behavior of the human organism and to foster the kind of personalities who can live in this new symbolic world and participate in a world community.

Perhaps the most astonishing development of recent times is the rise in the level of the aspirations of people who for generations have seemingly been content to accept ways of life that they now consider inadequate and frustrating. Thus people everywhere are faced with the task of renewing their culture and creating an industrial civilization. This means replacing the many self-defeating patterns that have been used for ordering, controlling, and rationalizing the non-rational human organism. The challenging creative task is to provide more fruitful and socially productive ways for coping with our persistent human predicament: each of us has to function as an organism while living as a personality in a symbolic cultural world and participating in the social order.

Evolution—Psychologists' Comments

Anatol Rapoport: What I have to say relates to the problem of plasticity and rigidity in man, in particular, his philosophical outlook. To all the definitions of man that have been offered since antiquity, I would like to add another (not very original, but perhaps it makes a point), a "dilemma-recognizing animal." In philosophy, this has been taken seriously to the extent that whole philosophical systems depend on the succession of dilemmas. A famous example is the Hegelian system of dialectics, which sees the world as a succession of theses, antitheses, and syntheses, in a spiraling process.

In particular, I should like to talk about the attitudes toward dilemmas which one finds in the history of science, as it is reflected in scientific philosophy. It is not an accident that the first dilemmas were pointed out by the sophists and the stoics, who were primarily concerned, not with natural philosophy, but with moral philosophy. These dilemmas are well known—Zeno's paradoxes, for example, the various antilogies, self-contradictions. It was not until much later that these dilemmas were finally resolved. The important thing is that these dilemmas were never resolved in the conceptual framework in which they arose, but only by a leap into another conceptual framework.

If we take one of Zeno's paradoxes (let us say the one about Achilles and the tortoise), it could not be resolved within the framework of strict Aristotelian logic in which it was posed; it could be resolved only with the invention of an entirely new concept, that of infinitesimal analysis. The dilemma of the Michelson-Morley experiment was an anomalous result that could not be explained on the basis of the prevailing metaphysics, that of the space-and-time framework of Newton. That could be resolved only by a leap into another kind of metaphysics, that of Einstein-Minkowsky.

Now this attitude toward the dilemma, which characterizes the man of science, the philosopher of science, is different from the conventional attitudes toward dilemmas. Conventionally, dilemmas have either led to polarized conflicts (I'm right; you're wrong), or to split-the-difference compromises, or else have been simply ignored—typically by experts. By definition an expert is a person who is not responsible for anything outside the field of his own competence. If a dilemma arises in any situation, the expert can say, "Look here, I'm only here to design this hydrogen bomb, and that's all."

COMMENTS

However, these ways of facing dilemmas are not the ways that have made the dilemmas such wonderfully creative situations in human affairs. It would be amusing to see what would happen if, for example, Zeno's paradox of Achilles and the tortoise were treated in one of the conventional ways. If it led to a polarized conflict, then you would have an empirically minded party maintaining that Achilles would certainly win the race, and a theoretically minded party maintaining with equal vigor that Achilles could not possibly win the race, that therefore he does not win it. If you had a compromise solution—which is the usual sort of thing in our political life, in the Anglo-Saxon tradition, with our balance-of-power arrangements, our lobbies, and our log-rolling, sometimes pointed out as the products of Anglo-Saxon genius—if that solution were applied to Zeno's paradox, then the race would probably be announced a tie. Or, one could say in a more sophisticated manner, who wins the race depends on one's frame of reference.

The most characteristically human creative thing about man's uniqueness is his ability to deal with dilemmas by getting *outside* of the dilemma situation by a radical re-examination of the evaluating process, of the language, of the metaphysics, of the philosophy, of the framework of assumptions—whatever has led to the dilemma in the first place. Instead of our trying to patch it up, we are sometimes given a beautiful opportunity to make an entirely different kind of formulation in which the dilemma becomes resolved, so that we go on to a new dilemma.

The dilemma being discussed at present is the problem of control versus freedom. The obvious and easy way to deal with it is by splitting into parties, groups, and partisan factions, or else, in a somewhat more polite way, by arriving at a sort of superficial compromise; but I submit here is an opportunity not to confine oneself to these kinds of solutions, but instead to seek a creative solution through an entirely new formulation.

Hallock Hoffman: The assumption underlying the present discussion (an assumption supported by all of us) is that men's behavior has changed as primitive societies have developed into the more complex associations of the modern world. This assumption parallels another, apparently agreed to, that man virtually stopped evolving biologically a good many thousand years ago. Genetic variations since that time have tended to be preserved by social environments.

Under these assumptions, we have now discussed the question of how men ought to design their environment so as to make themselves into the kind of creatures we believe they ought to become. As soon as the proposition is stated this bluntly, however, cries of distress ring out around the table. Carl Rogers seems certain that any self-conscious determination to make men over will make them worse; Anatol Rapoport wants to use the term "subversive" for the intention to reform men, so sure is he of the reaction to the proposal. Lawrence Kubie bridles at the word "control," equating it with brainwashing, suggesting that we as men have made no discernible improvement in ourselves in the past, and that we therefore ought to be doubtful about our ability to do so in the future. Erik Erikson treasures the personalities of individuals, and foresees nothing but catastrophe for us all if we adopt a self-conscious program of social improvement. Altogether, these objections lean heavily on the belief that men's minds are narrower than their experiences, with which no reasonable man could disagree; these conservatives conclude that man should decline to use the powers they now concede he has to shape himself by shaping his environment.

Against this argument, B. F. Skinner and Henry A. Murray, although starting from different interests and data, both claim that the power to shape men by influencing their environment exists, and that this power, without our appreciation of it, makes us what we are today.

In a curious way, I have seen the controversy between Skinner and Rogers—which I have followed in the journals over the last five or six years—as a needless controversy between two men who cannot learn to trust each other. It seems to me an example of the absence of the dyadic relationship Henry Murray holds up as the model for men if men are to have a future. "Control" for Rogers invokes images of men under other men's thumbs; "freedom" for Skinner raises specters of ancient superstitions that have aborted for centuries the scientific study of man. Each is correct, but not about the other; we must find ways to encourage the flowering of individual human beings in the myriad (but not infinite) systems of development open to them, and we must find ways to limit the uses men make of power over each other so that neither superstitious nor narrowly self-regarding men can injure others.

In looking at the present circumstance of men, I have come to regard certain features of our human environment as new, necessary in the future, and as yet uncomprehended in our general understanding. These features are:

1. Large populations in large social units. Though many men have lived under what appeared to be big governments, in nations having big populations, in times past, I believe the historic examples cannot accurately be compared to our present situation, for they were missing the present technology. A citizen of the Roman empire had some consciousness of the reach and inclusiveness of his social and political institutions, some of which, notably law, did become common to most parts of the empire. Now, however, transportation and communication make almost instantaneous our intercourse with vast territories and varied systems of ideas. No place is remote, and no idea really foreign. As a result, our politics seem empty. It is one characteristic of our age that we *know* that public pronouncements do not mean what they say. I believe this is not the result of any determination on the part of public leaders to tell lies, but rather of our inability to make truths big enough to cover what we are all exposed to.

2. The size of populations of going social units demands large-scale organization, which in turn requires specialization of function (bureaucracy and indirect control), representative systems for gathering political questions, deciding them, and disseminating the decisions. Until very recently, men have lived face-to-face with one another; the promulgation of an order affecting one man was done by, or in the name of, another man he could sometimes see. Suddenly, and in respect to an enormous number of affecting matters, the conditions of our lives are determined almost entirely by men we never see. Men have come to act upon one another in the manner of natural forces; this is what we mean by all the descriptions of people as "masses." We have constructed social institutions the objects of which are to make men interchangeable; and we have stumbled into a social relation the quality of which is to treat men as things. So far, we have not discovered how to extricate ourselves from this situa-

tion; for we want and even need the efficiency our social inventions make possible, and yet those inventions are making us into creatures we sense we dare not let ourselves become.

3. Along with the changes in magnitude of our institutions, we are accumulating technical information, and the technology and industrialization it brings, at an increasing rate. Until very recently, perhaps until fifty or seventy-five years ago, men were correct in believing that their grandsons could get along with the same ideas they had themselves. Now every man in his own lifetime must expect to change his own ideas and must add greatly to his information, merely to maintain his position in his social world. The rate of change has multiplied beyond our comprehension, and the demand for decisions has multiplied along with it. In Lincoln's generation, planners could wait for problems to arise, make plans to deal with them, and within reason anticipate that their plans would serve their children or perhaps several future generations. Now we must anticipate problems that have not yet arisen and solve them before we reach them, if our plans are to do us any good even for the moment.

These great new constraints upon mankind require that men themselves be made over. One immediate consequence of apprehending these constraints is that men are forced to think of their problems as world-wide, they are driven to recognize their common humanity, and they are encouraged to seek resolutions of their difficulties on a world scale. We have lately, at the Center for the Study of Democratic Institutions, become much involved with the idea of the world.

The idea of the world depends for its sustenance on the idea of law. What is meant by law is not merely statutes but the perception of order itself. To make a meaning is to make a law; or to make a law is to give a meaning to an otherwise disorderly situation. In its social form, law thus defined includes custom, constitutions, statutes, treaties, agreements, contracts, the common law, the rules of order and procedure—all the expressions of the idea of social order exhibited by society, to which the idea of order is central, all in operation through a complex social system of practical and judicial adjudication. Law is the record of general social decisions reached through a generality of participation.

The problem of these discussions has come to be, "How shall we make men into creatures who can continue to live together, now that they have learned how to kill one another so efficiently?" Because of the above ideas, I would suggest that this question could be phrased, what system of law can we construct to unify the world, to enable men to identify themselves with one another, to encourage them to discover their common interests, to force them to recognize their common humanity? That men's good is in some great dimension common, no one in this group doubts, and that men's present perspective obscures the commonalty of their good has been generally demonstrated. That men's actions are shaped by their ideas, that ideas and ideologies are a functional component of men's cultures, is the essence of Henry Murray's presentation.

In other words, I suggest that law is the instrument for reforming men—reminding you that "law" here refers not merely to statutes but to the whole array of social decisions that deal with and make visible the common good.

I would ask that we become law-makers, to make laws for men, that men may at last achieve civility.

Notes on the Authors

RALPH W. BURHOE, born in Somerville, Massachusetts, in 1911, is executive officer of the American Academy of Arts and Sciences. Previous to this he was on the administrative staff of the American Meteorological Society.

JAMES F. CROW, born in Phoenixville, Pennsylvania, in 1916, is professor of medical genetics at the University of Wisconsin. In 1960 he served as president of the Genetics Society of America, and currently is chairman of the Committee on Genetic Effects of Radiations of the National Academy of Science. In addition to his studies on the population genetics of Drosophila, he is the author of *Genetics Notes*.

HUDSON HOAGLAND, born in Rockaway, New Jersey, in 1899, is cofounder and Executive Director of the Worcester Foundation for Experimental Biology. He has taught at Boston, Cambridge (England), Clark, and Harvard Universities among others. Besides his serving in a long series of editing and administrative posts, he has contributed extensively to the literature on physiology, most recently as editor of *Hormones, Brain Function and Behavior*. His article, "Population Problems and the Control of Fertility," appeared in *Dædalus*, Summer 1959.

HERMANN J. MULLER, born in New York City in 1890, was Nobel Laureate in Physiology in 1946, and the recipient of the Newcomb Award of the AAAS in 1927 and the Kimber Genetics Award in 1955. From 1933-1937 he was senior geneticist at the Institute of Genetics in Moscow, and from 1937-1940 a research associate and lecturer at the Institute of Animal Genetics in Edinburgh; since 1945 he has been professor of zoology at Indiana University. He has been president of the American Humanist Association, United States delegate at the Geneva conference, "Atoms for Peace" (1958), and participant in the second and fourth Pugwash Conferences. His publications include: *Out of the Night: A Biologist's View of the Future;* and *Genetics, Medicine, and Man*.

HENRY A. MURRAY, born in New York City in 1893, is professor of clinical psychology at Harvard University. His principal field of interest has been the development and assessment of personality. His publications include *Explorations in Personality* and *Assessment of Men*, the latter prepared and edited during his service with the O.S.S. in World War II. His most recent contribution to *Dædalus* was in the issue "Myth and Mythmaking" (Spring 1959).

WALTER A. ROSENBLITH, born in Vienna in 1913, is professor of communications biophysics and acting chairman of the Center for Communication Sciences at the Massachusetts Institute of Technology. Since graduating from the Ecole Supérieure d'Electricité (Paris, 1937), he has contributed extensively to various periodicals and compilations on the quantification of electrical activity in the nervous system and on the handling of sensory information by organisms. He is presently chairman of the Committee on the Use of Electronic Computers in the Life Sciences (National Academy of Sciences-National Research Council) and member of the Central Committee of the International Brain Research Organization. He is the author of *Noise and Man* (with Kenneth N. Stevens) and editor of *Sensory Communication* (forthcoming).

DEMITRI B. SHIMKIN, born in Omsk, Russia, in 1916, is professor of anthropology at the University of Illinois. He has been lecturer in anthropology at the Russian Research Center at Harvard University (1948-1953), social science analyst for the United States Bureau of the Census (1953), and lecturer at George Washington University (1955). He is the author of *Minerals, a Key to Soviet Power*.

B. F. SKINNER, born in Susquehanna, Pennsylvania, in 1904, is Edgar Pierce Professor of Psychology at Harvard University. The recipient of the Warren Medal of the Society of Experimental Psychology, and the Distinguished Scientific Contribution Award of the American Psychological Association, he is the author of: *Behavior of Organisms; Walden Two; Science and Human Behavior; Verbal Behavior; Schedules of Reinforcement* (with C. B. Ferster); and *Cumulative Record*.

JULIAN H. STEWARD, born in Washington, D.C., in 1902, is professor of anthropology at the University of Illinois. After serving as senior anthropologist at the Bureau of American Ethnology, he became director of the Institute of Social Anthropology at the Smithsonian Institution and later professor of anthropology at Columbia University. His publications include: *Theory of Culture Change; People of Puerto Rico* (with others); *Native People of South America* (with L. C. Faron); and (as editor), *Handbook of South American Indians* (6 vols.).

Glossary

Dysgenic: a character or genetic factor which lowers the fitness of an individual or a population.

Darwinian fitness: the contribution to the gene pool of the next generation.

Gene locus: the location on a chromosome where a given genetic factor is localized.

Gene migration: the movement of genes (through interbreeding) from one population to others.

Gene pools: the totality of the genes of a given population existing at a given time.

Genotype: the totality of genetic factors which control a character or the characters of an individual.

Heterozygous: an individual with different alternate genetic factors (alleles) at the homologous (corresponding) loci of the two (parental) chromosomes.

Homozygous: an individual with identical alternate genetic factors (alleles) at the homologous (corresponding) loci of the two (parental) chromosomes.

Inbred: the interbreeding of closely related, hence genetically similar, individuals.

Infraspecific populations: populations belonging to the same species.

Interspecific isolation: the nonoccurrence of crossbreeding between species.

Intraspecific selection: natural selection among individuals of the same species.

Irreversibility: the theory that a given structure or adaptation which has been lost in evolution cannot be restored exactly to its prior condition.

Meiosis: special cell divisions in the developing sperm cells which result in a reduction of chromosome number. (During meiosis an exchange of sections may occur between homologous chromosomes.)

Ontogenetic: dealing with the development, particularly the embryo genesis of the individual.

Orthogenesis: an erroneous assumption that evolution follows a predetermined rectilinear pathway.

Outbred: the interbreeding of unrelated, hence usually genetically dissimilar, individuals.

Parthenogenesis: the development of an egg without fertilization.

Phenotype: the totality of characteristics (attributes) of an individual.

Phylogenetic: dealing with the changes in an evolutionary line.

Spermatogonia: cells in the testes which through division and modification give rise to the spermatozoa.

Subspeciation: the development of well-defined geographic races (subspecies) within a species.

Supraspecific evolution: evolution above the species level, evolution of higher categories and major trends.

Stirps: stock, race, family, lineage.

Synapses: in neurology, the contact between two neurons (nerve cells); in cytology, the pairing of homologous chromosomes during meiosis.

Zygote: the cell (individual) which results from the fertilization of an egg cell, the fertilized egg.

Conferences on Evolutionary Theory and Human Progress

CONFERENCE A: 30 September–2 October 1960

Genetics and the Direction of Human Evolution

Roster of Conferees: Ralph W. Burhoe, James F. Crow, Theodosius Dobzhansky, Leonard Engel, Donald H. Fleming, Lawrence K. Frank, Charles Frankel, John L. Fuller, Ralph W. Gerard, H. Bentley Glass, Stephen R. Graubard, Garrett Hardin, Hudson Hoagland, Mahlon B. Hoagland, John C. Honey, I. Michael Lerner, Kirtley F. Mather, Ernst Mayr, Robert P. Morison, Hermann J. Muller, Henry A. Murray, Ernest Nagel, Theodore T. Puck, Curt P. Richter, Walter A. Rosenblith, J. Paul Scott, George G. Simpson, B. F. Skinner, and James N. Spuhler.

CONFERENCE B: 4–6 November 1960

The Dynamics and Direction of Social Evolution

Roster of Conferees: J. O. Brew, Robert J. Braidwood, Ralph W. Burhoe, James F. Crow, Theodosius Dobzhansky, Leonard Engel, Donald H. Fleming, Lawrence K. Frank, Morton H. Fried, Ralph W. Gerard, Irving Goldman, Ward H. Goodenough, Stephen R. Graubard, Garrett Hardin, Hudson Hoagland, John C. Honey, Alex Inkeles, Harold D. Lasswell, Chauncey D. Leake, I. Michael Lerner, Alexander Lesser, Robert A. Manners, Ernst Mayr, Margaret Mead, Hermann J. Muller, Henry A. Murray, Talcott Parsons, Van R. Potter, Anne Roe, Walter A. Rosenblith, J. Paul Scott, Michael Scriven, Demitri B. Shimkin, George G. Simpson, B. F. Skinner, Julian H. Steward (in absentia), and Sherwood L. Washburn.

CONFERENCE C: 2–4 December 1960

Evolution and the Individual

Roster of Conferees: Carl A. Binger, Ralph W. Burhoe, Leonard Engel, Erik H. Erikson, Clarence H. Faust, Donald H. Fleming, Lawrence K. Frank, Ralph W. Gerard, Stephen R. Graubard, Harry Harlow, Hudson Hoagland, Hallock Hoffman, John C. Honey, Lawrence S. Kubie, Harold D. Lasswell, Donald G. Marquis, Abraham H. Maslow, Kirtley F. Mather, Ernst Mayr, Hermann J. Muller, Henry A. Murray, Anatol Rapoport, Carl R. Rogers, Walter A. Rosenblith, J. Paul Scott, Michael Scriven, and B. F. Skinner.

Bei Fragen zur Produktsicherheit wenden Sie sich bitte an:
If you have any questions regarding product safety,
please contact:

Walter de Gruyter GmbH
Genthiner Straße 13
10785 Berlin
productsafety@degruyterbrill.com